Egg Laying Characteristics of the Hen

by Professor James Dryden

with an introduction by Jackson Chambers

IMPORTANT NOTE & DISCLAIMER

IMPORTANT NOTE :

As with all reprinted books of this age that are intended to perfectly reproduce the original edition, considerable pains and effort had to be undertaken to correct fading and sometimes outright damage to existing proofs of this title.

At times, this task can be quite monumental, requiring an almost total rebuilding of some pages from digital proofs of multiple copies. Despite this, imperfections still sometimes exist in the final proof and may detract slightly from the visual appearance of the text.

Some images may suffer from reduced quality due to anomalies in the original scan.

DISCLAIMER :

Due to the age of this book, some methods or practices may have been deemed unsafe or unacceptable in the interim years. In utilizing the information herein, you do so at your own risk.

We republish antiquarian books with no judgment or revisionism, solely for their historical and cultural importance, and for educational purposes.

Self Reliance Books

Get more historic titles on animal and stock breeding, gardening and old fashioned skills by visiting us at:

http://selfreliancebooks.blogspot.com/

introduction

Here at **Self-Reliance Books** we are dedicated to bringing you the best in *dusty-old-book-knowledge* – this time, an old book on the production comparisons of egg-laying Hens.

This special edition of ***Egg Laying Characteristics of the Hen*** was written by Professor James Dryden, and first published in 1821, making it just shy of a century old.

The book contains sections on *Problems Outlined, Results With Barred Plymouth Rocks, Results with White Leghorns, Crossing and Inheritance, Effect of Close Confinement,* and more.

This is a fantastic old book and is an essential addition to the libraries of all Poultry breeders and egg producers.

~ *Jackson Chambers*

State of Jefferson, April 2018

2

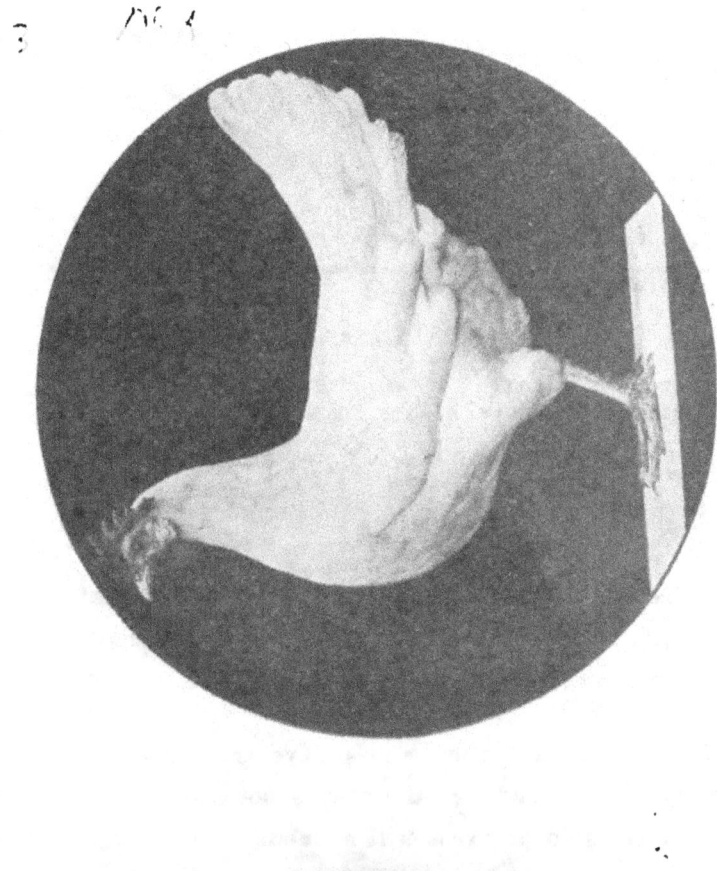

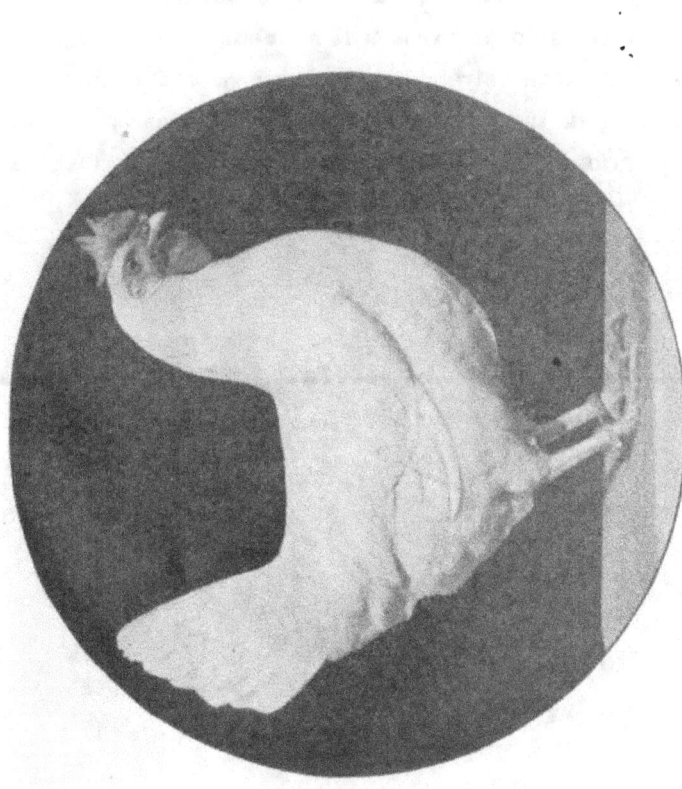

Fig. 1. At left, C521 (Lady MacDuff) representing the Station's "Oregons." Her record was 303. At right, hen A27, representing the Station's White Leghorns; the first hen in the world to make trap-nest record of 1000 eggs in lifetime. Her complete record was 1188 eggs. All the Station's Oregons and Leghorns have more or less of the blood of these two hens, which have thousands of descendants throughout the country. The first hen in the world with trap-nest record of 300 eggs in a year.

The main conclusions drawn from the experiments reported in this bulletin are:

1. High fecundity is inherited.

2. Selection of breeding stock on the basis of annual trap-nest records regardless of prepotency or tested qualities, is a certain method of increasing egg production.

3. Some hens and some males have the power of transmitting high fecundity; others have not this power. Therefore more rapid progress will be made in increasing the production of the strain if only those hens and those males be used in the breeding pens that have shown by the egg records of their pullets, or by the progeny test, that they possess the power of transmitting high egg production.

Egg-Laying Characteristics of the Hen

INTRODUCTION

When poultry breeding experiments were started at this Station, twelve years ago, there was considerable discussion as to the effect of selective breeding in increasing egg production. Doubt was frequently expressed as to the possibility of increasing production by selection of breeding stock on the basis of trap-nest records of production. There had been no sufficient demonstration of the fact of transmission of high egg-laying qualities. There was some evidence that the contrary was true. At the Maine station, after nine years of experiments the results as reported were largely negative in character. There had been apparently no increase in production. Beyond that there was little literature on the subject embodying conclusive data.

It was agreed that the subject required further investigation. It was highly important from the standpoint of the poultry producer, as well as from the standpoint of the consumer of poultry products, that production be increased. The costs of egg production were high, and unless they could be reduced, poultry keeping would continue to be a business of hazard.

Previous experiments by the writer had shown a wide variation in individual production, and if it should prove that those variations were inherited, it would seem then, that proper breeding, by increasing production, would be of great importance in lowering the costs.

At that time the flocks of the United States as an average laid fewer than seventy-five eggs a year per hen, while trap-nest records showed yields exceeding two hundred a hen and others less than two dozen a hen, even within the same breed and under similar conditions of care. This fact indicated great possibilities for the poultry industry, should it be proved that those variations were inherited. The production of the hens of the United States as given is based on census figures; these apply to hens of all breeds and different ages and would not be fully comparable with the production of pullets, or the first year's production.

Until the invention of the trap-nest, poultry breeding, so far as egg production was concerned, was an uncertain business because the poultryman had no means of knowing which were the best layers in his flock. With the trap-nest it was then possible to attack the problem of increasing production in a practical, scientific way.

Our breeding work began in the fall of 1908 and has continued without intermission up until this time, though the scope of investigations has been necessarily limited. With the funds and help available, it has not been possible to cover as many closely related problems as was desired. Breeding work is necessarily slow, and often disappointing because of the variety of conditions under which the flocks are kept during the long period of experiment. Egg production fluctuates greatly as the environmental conditions vary. Changes in the weather affect the egg yield, and it is impossible so to regulate other conditions that the yield is not appreciably affected by sudden temperature and other changes in the weather. Where the results depend upon comparisons extending over several years, with flocks of different generations, this is a disturbing element that can possibly be overcome only by long-continued, carefully conducted experiments.

While the feed rations and methods of feeding can be fairly well controlled, there will be unavoidable differences. It is not always possible to get the same quality of feed, year after year. Another limiting factor is diseases. To maintain a strict quarantine so that the flocks of one year will show as great freedom from diseases as those of another, is possible only in part. Change in the attendant who feeds, tends trapnests, and does other necessary work in the poultry yards, also has its influence on production. Change of attendant is sometimes unavoidable.

Another factor, probably the most difficult to control of any, is that of the hatching and rearing of the chicks. The vigor of the chicks is influenced by the care and feeding of the breeding stock, by the methods of hatching, and by the brooding and feeding of the chicks. Any difference in the vigor of the hen will be reflected in the egg yield. The date of hatching, or the date at which the pullet comes to laying maturity, is another factor that has a decided influence on the first year's production.

It is thus impossible so to control all these and other environmental conditions that we can know with mathematical definiteness that the egg producion of one particular generation of fowls for any stated period measures with absolute certainty the laying capacity of the flock.

Nor is it essential that the environmental factors actually be the same year after year to prove by the egg records for the year the transmission of high fecundity. With a sufficient number of individuals, and a sufficient number of generations, the fact will be proved if there is a consistent and marked increase, even though the conditions may and must vary, year after year.

Scientific demonstration need not and can not depend upon mathematical and mechanical control of environmental factors. At the same time it must be conceded that the degree of success in controlling these conditions will measure very largely the success of the breeding experiment.

PROBLEMS OUTLINED

The general subject of investigation is the inheritance of egg-laying characteristics of the hen, covering the following related problems:

(1) Is high fecundity inherited?

(2) Relation of short-period production, or rate of laying, to the annual and biennial record of the hen.

(3) Does the first year's production indicate the capacity of the hen in subsequent years?

(4) Will selection on the basis of high annual or short-period production increase the profitable life of the hen?

(5) Correlation between early laying maturity of the pullet and high fecundity.

(6) Correlation between late laying and high fecundity.

(7) The effect of inbreeding on vigor and productive qualities.

(8) The relative influence of sire and dam on the egg production of the offspring.

(9) Correlation between type of hen and egg production.

(10) Correlation between rate of laying and composition of the egg.

This bulletin is confined to a study of our breeding records bearing on the first four problems enumerated above. Later bulletins will report work on the other problems.

There is also included in this bulletin a report of a preliminary nature on the tenth problem, the correlation between rate of laying and composition of the egg.

THE PLAN

For the experiments, Barred Plymouth Rocks and White Leghorns were chosen, the former because they are a variety of an American breed, very popular on the farms for general purposes, the latter because they are the most popular breed on commercial egg farms. A cross between these breeds was made, as a check on the others, because there is a somewhat widespread conviction that some of the pure-breds through close breeding have lost more or less vigor, and that a cross will restore the vigor. If therefore, the cross gave a better egg yield than the pure-breds it would be strong circumstantial evidence that the pure-breds lacked in vigor. Then, if after several generations the cross-bred birds showed an increase in production, while a similar increase was shown in the pure-breds it would be strong evidence that the breeding was effective, and that high fecundity was inherited. On the other hand, without the cross-bred check it might be questionable, whether the increase in the pure-bred was not due solely to increased vigor, which might be secured from proper mating and breeding. It is known that vigor and production are closely related.

IS HIGH FECUNDITY INHERITED?

A flock of 113 Barred Plymouth Rock pullets were purchased from six different breeders in Oregon, and 63 White Leghorn pullets from several Oregon poultrymen. These pullets were put into their houses early in the fall of 1908. The houses used were new, portable, open-front, or colony houses, eight by twelve feet in size. The fowls had the liberty of roomy yards at all seasons. The hen doors in the houses were open at all times. The double-yard system was used and the houses were moved once a year onto clean ground. The same style of house and the same system of yarding were used in all the years of the experiment. The same system of feeding was followed, as nearly as possible, throughout the experiment. Inbreeding was avoided, and practically no new blood was added. In the last three or four years, however, especially in the case of the Leghorn, the blood lines had become closely drawn, so that in the case of the Leghorn the results of the later years are not properly comparable with those of former years.

It was aimed to use each year in experiments only pullets that were mature and about ready to lay by November 1, so that the egg records of different years might afford a fair basis of comparison. This practice, however, could not be followed arbitrarily. It was not always possible to hatch them at the same time, and where the purpose was to hatch as many chicks from certain hens as possible, the date of hatching must vary more or less. Any of the yards or hens that were used in our comparisons, however, did not vary greatly in date of laying maturity. On this account and for other reasons, the number of pullets used in the tabulation for some of the years is rather small. It is believed, however, that enough individuals have been used to make it possible to form conclusions on the subject of investigation.

The capacity of a hen is commonly measured by her annual egg yield, and the proof of transmission is shown if the female progeny produces not necessarily as many eggs as the parent, but a certain increase over the average production of the unbred flock or breed. It may be true, however, that her real capacity is better shown in shorter periods of production, because in shorter periods, environmental conditions are under better control. Our records, therefore, will be studied from annual records, as well as longer and shorter periods.

Artificial lighting was not used in these experiments. It had not come into use when the experiments were started, and to have adopted it in later years would have introduced a factor that would render impossible proper interpretation of results of different generations, if some were under artificial lights and some were not.

In all experiments the first year begins on the date the first egg was laid, and ends twelve months from that date. The second year begins at the end of the first laying year. In all cases, hens that died before the year was completed are not included, except that a record of number died is given.

Eggs laid on the floor or outside of the trap-nests are added to the total production of the flock and show in the yard average, but not in the individual-hen records.

RESULTS WITH BARRED PLYMOUTH ROCKS

First Generation or Foundation Stock. In the case of Barred Plymouth Rocks ninety-two hens are used in the experiment relating to inheritance of fecundity. For these there is a complete record for the first year's production.

The first and most significant thing in the production of all the years is the wide variation in the records of individual hens. This applies to the two different breeds, as well as the crosses. In the original flock of ninety-two Barred Rocks this variation, as shown in Table 1, extends from 218 eggs as the highest to 6 as the lowest, the average of all being 82.67, or including eggs laid outside of trap-nests, 86.14 eggs. In other words, between the highest and lowest record there is a range of 212 eggs. It will be noted that there is a gradual drop all of the way down from the highest to the lowest. A column is added giving the best two-months record of each hen. The significance of this record will be discussed on a later page.

Second Generation. The production in the second year, 1909-10, is given in Table II, for yard 6, 28 hens. These were hatched from the first-year flocks but without selection, and sired by males of pedigree of unknown production. In this case the highest record was 183 and the lowest 60, a range between highest and lowest of 123. The yard possibly would represent a better average of the production of strain or strains than the original flock, or rather make a better starting point in studying the effect of breeding, because the pullets of this yard were raised on the Station plant, and so far as the hatching and rearing are

TABLE I. ANNUAL AND BEST TWO-MONTHS PRODUCTION
Yards 4 and 5, 1908-09, Barred Plymouth Rocks.

Hen No.	Production 1st year	Best 2 months	Hen No.	Production 1st year	Best 2 months	Hen No.	Production 1st year	Best 2 months
1....	36	10	35....	66	37	78....	68	30
2....	93	28	36....	70	30	80....	71	20
3....	105	32	37....	47	18	81....	39	29
4....	124	26	40....	183	47	82....	61	29
5....	131	44	41....	118	45	83....	45	22
6....	80	27	43....	105	44	84....	56	23
7....	47	19	44....	59	28	86....	106	34
8....	67	28	45....	140	46	87....	64	30
9....	51	25	47....	82	36	88....	82	33
10....	88	28	48....	47	29	89....	153	35
11....	31	15	50....	91	45	90....	83	43
12....	72	27	51....	115	41	91....	49	22
13....	71	28	52....	70	26	93....	80	32
15....	155	38	53....	83	45	95....	134	41
17....	51	27	54....	112	46	96....	64	35
18....	33	14	55....	57	28	97....	100	32
19....	73	32	56....	141	42	98....	81	29
20....	65	24	57....	59	31	99....	167	48
21....	57	28	58....	37	22	201....	52	30
22....	6	3	61....	85	31	202....	103	37
23....	74	31	62....	106	31	203....	96	44
25....	69	28	64....	49	23	204....	101	36
26....	74	38	65....	218	50	205....	59	23
27....	68	31	66....	74	33	207....	110	47
28....	37	20	67....	86	28	208....	52	26
29....	42	22	68....	78	24	209....	53	21
30....	84	41	70....	133	36	210....	58	30
31....	93	33	72....	67	25	211....	160	48
32....	136	40	73....	66	24	212....	46	34
33....	131	37	74....	76	32	213....	84	36
34....	76	25	77....	89	28			
Average of yard							82.67	31.29
Unidentified eggs added							86.14	

TABLE II. ANNUAL AND BEST TWO-MONTHS PRODUCTION
Yard 6, 1909-10, Barred Plymouth Rocks.

Hen No.	Production 1st year	Best 2 months	Hen No.	Production 1st year	Best 2 months	Hen No.	Production 1st year	Best 2 months
201....	104	33	211....	167	43	221....	103	32
202....	128	41	212....	138	30	222....	140	44
203....	114	31	213....	144	37	223....	60	32
204....	183	40	214....	115	37	224....	138	43
205....	88	28	215....	134	31	225....	102	33
206....	85	37	216....	104	34	226....	160	38
207....	169	43	218....	76	24	227....	76	28
208....	46	25	219....	88	27	228....	104	28
209....	179	33	220....	110	25	229....	75	29
210....	134	38						
Average of yard							117.64	33.71
Unidentified eggs added							120.68	

TABLE III. ANNUAL AND BEST TWO-MONTHS PRODUCTION
Yard 15, 1910-11, Barred Plymouth Rocks.

Hen No.	Production 1st year	Best 2 months	Hen No.	Production 1st year	Best 2 months	Hen No	Production 1st year	Best 2 months
A 75	158	40	A 92	144	38	A 106	202	44
A 76	183	43	A 93	201	42	A 107	122	33
A 77	214	50	A 94	20	8	A 108	166	37
A 78	215	44	A 95	128	31	A 109	179	41
A 79	219	50	A 96	204	43	A 110	175	35
A 80	168	40	A 97	79	26	A 111	204	53
A 81	203	42	A 98	130	35	A 113	175	37
A 83	167	37	A 99	134	33	A 115	188	39
A 84	44	28	A 100	125	42	A 116	218	50
A 85	182	43	A 101	137	37	A 118	191	45
A 86	133	35	A 102	169	39	A 119	156	42
A 87	6	6	A 103	168	37	A 120	148	41
A 88	187	45	A 104	158	35	A 121	168	39
A 91	158	36	A 105	200	46	A 122	259	49

Average of yard 161.78   38.48

Unidentified eggs added 164.28

TABLE IV. ANNUAL AND BEST TWO-MONTHS PRODUCTION
Yard 6, 1912-13, Barred Plymouth Rocks.

Hen No.	Production 1st year	Best 2 months	Dam	Eggs 1st yr.	D.D.	Eggs 1st yr.	S.D.	Eggs 1st yr.
C 32	188	50	A 78	213	....	....	207	169
C 33	139	36	A 111	204	....	....	65	218
C 34	180	42	A 81	203	....	....	207	169
C 35	225	53	A 79	219	....	....	207	169
C 36	89	34	A 81	203	....	....	207	169
C 37	109	47	A 79	219	....	....	207	169
C 38	213	49	A 111	204	....	....	65	218
C 39	145	49	A 111	204	....	....	65	218
C 40	151	39	A 77	214	....	....	65	218
C 41	177	45	A 79	219	....	....	207	169
C 42	138	45	A 111	204	....	....	65	218
C 43	142	41	A 81	203	....	....	207	169
C 44	227	49	A 111	204	....	....	65	218
C 45	171	46	A 81	203	....	....	207	169
C 46	201	41	A 77	214	....	....	65	218
C 47	156	41	A 81	203	....	....	207	169
C 48	268	51	A 111	204	....	....	65	218
C 49	194	44	A 79	219	....	....	207	169
C 50	224	53	A 77	214	....	....	65	218
C 51	196	56	A 111	204	....	....	65	218
C 52	157	52	A 77	214	....	....	65	218
C 53	144	43	A 81	203	....	....	207	169
C 54	186	49	A 79	219	....	....	207	169
C 55	142	47	A 111	204	....	....	65	218
C 56	210	50	A 111	204	....	....	65	218
C 57	178	44	A 77	214	....	....	65	218
C 58	205	55	A 77	214	....	....	65	218
C 59	209	41	A 79	219	....	....	207	169
C 60	3	2	A 111	204	....	....	65	218
C 62	175	45	A 78	213	....	....	207	169
C 63	140	38	A 77	214	....	....	65	218
C 64	235	50	A 122	259	....	....	207	169
C 65	123	39	A 81	203	....	....	207	169
C 67	187	44	A 111	204	....	....	65	218
C 68	147	45	A 78	213	....	....	207	169
C 69	232	55	A 77	214	....	....	65	218
C 70	152	50	A 77	214	....	....	65	218
C 149	171	40	A 78	213	....	....	207	169
①	174.13	44.74		210.94				194.79
②	178.16							

① Average of yard. ② Unidentified eggs added.

TABLE V. ANNUAL AND BEST TWO-MONTHS PRODUCTION
Yard 7, 1912-13, Barred Plymouth Rocks.

Hen No.	Production 1st year	Best 2 months	Dam	Eggs 1st yr.	D.D.	Eggs 1st yr.	S.D.	Eggs 1st yr.
C 72	155	43	207	169	....	....	65	218
C 73	207	51	204	183	....	....	65	218
C 75	172	43	207	169	....	....	65	218
C 76	198	42	B 154	191	204	183	207	169
C 77	163	42	....	....	....	....	....	....
C 78	199	45	A 116	218	65	218	207	169
C 79	201	46	207	169	....	....	65	218
C 80	202	51	....	....	....	....	....	....
C 81	180	39	B 150	105	A 116	218	207	169
C 82	200	43	....	....	....	....	....	....
C 83	186	49	A 106	202	....	....	207	169
C 84	165	41	B 154	191	204	183	207	169
C 85	188	44	226	160	....	....	65	218
C 86	176	44	A 106	202	....	....	207	169
C 87	204	44	B 150	105	A 116	218	207	169
C 88	188	50	226	160	....	....	65	218
C 89	215	49	207	169	....	....	65	218
C 90	215	51	B 154	191	204	183	207	169
C 91	119	35	....	....	....	....	....	....
C 92	137	48	A 88	187	....	....	....	....
C 93	161	45	....	....	....	....	....	....
C 94	175	46	....	....	....	....	....	....
C 96	180	49	207	169	....	....	65	218
C 97	149	33	....	....	....	....	....	....
C 99	179	49	A 96	204	....	....	207	169
C 100	111	39	A 105	200	....	....	207	169
C 101	125	41	A 116	218	65	218	207	169
C 102	111	37	A 105	200	....	....	207	169
C 103	193	46	A 111	204	....	....	65	218
C 104	174	43	....	....	....	....	....	....
C 105	188	48	....	....	....	....	....	....
C 106	189	48	A 106	202	....	....	207	169
C 107	133	46	A 105	200	....	....	207	169
C 108	193	50	B 151	136	A 116	218	207	169
C 110	177	50	....	....	....	....	....	....
C 111	178	50	A 106	202	....	....	207	169
C 112	219	53	B 154	191	204	183	207	169
①......	177.43	45.22		181.37		202.44		185.96
②......	180.97							

① Average of yard. ② Unidentified eggs added.

concerned, under the same conditions as those of subsequent years. The average of the annual individual records of this yard was 117.64 a hen, or including eggs laid outside of nest, 120.68.

Third Generation. Table III gives record for yard 15, pullets of the second generation from the original flock. These pullets were not pedigreed; that is, the egg record of dam and sire's dam is not known except in the case of two hens; but the pullets were taken from the original flock as a whole after one-third of the poorest had been discarded. Of the two hens mentioned, one laid 218 eggs and her dam laid also 218; the other was from a poor layer, laying 45 eggs.

The record of this flock is higher than would be expected from the production of the dams, but the explanation is, no doubt, in the possibility that environmental conditions were specially favorable to high production. The record shows that this was true in some respects at least. They were hatched at a uniform time, and came to laying maturity early in the fall; in this respect their conditions were more favorable than was the case in earlier or later years.

The significance of this record is discussed later in the study of rate of laying or short-period production. The fact, however, that the flock from which yard-15 pullets were produced was a culled flock, approximately one-third of the poor layers having been taken out, is no doubt responsible for some of this increase.

Fourth Generation. The records of the pullet flocks of year 1912-13, Tables IV, V, VI, offer the first results of pedigree work, or the records of pullets of known production pedigree on both dam and sire side. The 38 pullets shown in Table IV were from the six best hens in yard 15, 1910-11, whose records are reported in Table III, and whose sires were two males from dams shown in Tables I and II; namely, 65 and 207, records 218 and 169 respectively. Table IV shows that the 38 pullets varied in production from 268 eggs to 3 eggs, and averaged 178.16 eggs. The average of their dams was 210.94 and of the sire's dam 194.79. It is worthy of note that the highest and lowest records in this yard were made by two full sisters, though probably no great importance can be placed upon the record of the lowest hen, C60, which was 3 eggs. While this hen was in apparently good condition until Oct. 19, 1915, near the end of the third year, when she was marketed, it is not impossible

TABLE VI.   ANNUAL AND BEST TWO-MONTHS PRODUCTION
Yard 8, 1912-13, Barred Plymouth Rocks.

Hen No.	Production 1st year	Best 2 months	Dam	Eggs 1st yr.	D.D.	Eggs 1st yr	S.D.	Eggs 1st yr
C 115	227	52	B 151	136	....	....	207	169
C 116	139	35	A 96	204	....	....	207	169
C 117	186	48	A 79	219	....	....	207	169
C 118	173	49	207	169	....	....	65	218
C 119	241	50	A 96	204	....	....	207	169
C 120	173	41	A 81	203	....	....	207	169
C 121	166	36	A 106	202	....	....	207	169
C 122	183	44	A 79	219	....	....	207	169
C 123	139	42	A 78	213	....	....	207	169
C 124	149	46	A 78	213	....	....	207	169
C 125	158	40	A 79	219	....	....	207	169
C 126	107	43	A 116	218	....	....	207	169
C 127	161	50	A 111	204	....	....	65	218
C 128	196	58	A 77	214	....	....	65	218
C 129	174	47	A 79	219	....	....	207	169
C 130	193	41	A 95	201	....	....	207	169
C 133	207	40	207	169	....	....	65	218
C 134	182	55	A 115	188	....	....	207	169
C 136	191	42	B 150	105	....	....	207	169
C 137	223	48	A 111	204	....	....	65	218
C 140	126	43	226	160	....	....	65	218
C 141	144	45	A 78	213	....	....	207	169
C 143	130	42	204	183	....	....	65	218
C 145	190	51	A 79	219	....	....	207	169
C 146	209	52	A 122	259	....	....	207	169
C 147	177	48	A 79	219	....	....	207	169
C 148	166	36	A 81	203	....	....	207	169
C 150	125	39	A 116	218	....	....	207	169
C 151	107	27	A 77	214	....	....	65	218
C 152	195	52	207	169	....	....	65	218
C 154	139	42	A 96	204	....	....	207	169
C 156	141	40	A 111	204	....	....	65	218
C 157	189	38	B 150	105	....	....	207	169
①......	171.69	44.30	......	196.76	....	....	....	183.85
②......	174.18							

① Average of yard.   ② Unidentified eggs added.

TABLE VII. ANNUAL AND BEST TWO-MONTHS PRODUCTION
Yard B, 1913-14, Barred Plymouth Rocks.

Hen No.	Production 1st year	Best 2 months	Dam	Eggs 1st yr.	D.D.	Eggs 1st yr.	S.D.	Eggs 1st yr.
D 66	150	32	C 69	232	A 77	214	A 116	218
D 67	151	51	C 33	139	A 111	204	A 78	213
D 68	219	55	Yd. 5-6	......	....	....	....	....
D 69	229	54	A 79	219	....	....	A 122	259
D 70	144	42	C 61	158	A 79	219	65	218
D 71	209	55	C 64	235	A 122	259	A 116	218
D 73	165	43	H 26 N	139		204	183	A 81	203
D 74	244	54	A 77	214	....	... .	A 122	259
D 75	201	46	A 77	214	....	....	A 122	259
D 77	208	48	C 64	235	A 122	259	A 116	218
D 78	215	48	A 79	219	....	....	A 122	259
D 79	109	47	226	160	65	218	A 81	203
D 80	98	44	C 58	205	A 77	214	A 116	218
D 83	190	48	C 64	235	A 122	259	A 116	218
D 84	236	53	C 146	209	A 122	259	A 77	214
D 85	161	47	A 105	200	....	....	A 81	203
D 87	243	52	C 32	188	A 78	213	A 116	218
D 88	153	44	B 159	188	207	169	A 77	214
D 89	151	42	C 43	142	A 81	203	A 116	218
D 90	240	54	C 64	235	A 122	259	A 116	218
D 91	147	37	C 149	171	A 78	213	A 116	218
D 92	259	53	C 35	225	A 79	219	65	218
D 93	171	54	C 69	232	A 77	214	A 116	218
D 94	112	42	A 93	201	....	....	A 81	203
D 95	194	48	C 35	225	A 79	219	65	218
D 97	201	51	Yd. 6	......	....	....	....	....
D 98	182	49	C 69	232	A 77	214	A 116	218
D 99	86	35	C 40	151	A 77	214	A 116	218
D 100	213	46	C 44	227	A 111	204	A 78	213
D 101	166	50	C 45	171	A 81	203	A 116	218
D 102	166	33	204	183	....	....	A 77	214
D 103	179	46	226	160	65	218	A 81	203
D 104	183	48	A 79	219	....	....	A 122	259
D 105	167	54	C 46	201	A 77	214	A 116	218
D 106	225	49	C 50	224	A 77	214	A 116	218
D 107	179	43	C 52	157	A 77	214	A 116	218
D 108	187	48	C 41	177	A 79	219	65	218
D 109	130	45	A 106	202	....	....	A 122	259
D 111	176	52	C 37	199	A 79	219	65	218
D 112	160	49	C 58	205	A 77	214	A 116	218
D 113	199	51	A 93	201	....	....	A 81	203
D 115	170	43	B 246	116	226	160	A 77	214
D 116	184	50	B 159	188	207	169	A 77	214
D 117	169	47	204	183	....	....	A 77	214
D 118	233	55	C 64	235	A 122	259	A 116	218
D 119	209	50	C 64	235	A 122	259	A 116	218
D 120	95	41	C 37	199	A 79	219	65	218
D 121	149	43	B 160	119	207	169	A 77	214
D 122	162	37	B 246	116	226	160	A 77	214
D 123	233	52	A 79	219	....	....	A 122	259
D 124	116	37	226	160	65	218	A 81	203
D 126	177	41	Yd. 7	......	....	....	A 111	204
①......	178.75	46.88	....	193.86	....	215.05	....	220.52
②......	185.00							

① Average of yard.   ② Unidentified eggs added.

that her low performance was due to some physical condition of the egg-laying organs, not at all related to any inherited capacity to lay.

Table V gives a record of yard 7 containing 37 pullets that made complete one-year records.. They averaged 180.97 eggs a hen. Their dams averaged in their first year 181.37 eggs, and their sire's dam 185.96 eggs.

TABLE VIII. ANNUAL AND BEST TWO-MONTHS PRODUCTION
Yard C. 1913-14, Barred Plymouth Rocks.

Hen No.	Production 1st year	Production Best 2 months	Dam	Eggs 1st yr.	D.D.	Eggs 1st yr.	S.D.	Eggs 1st yr.
D 128	145	43	A 78	213	......	......	A 122	259
D 129	189	49	C 68	147	A 78	213	A 116	218
D 130	183	48	C 70	152	A 77	214	A 116	218
D 131	173	46	A 78	213	......	......	A 122	259
D 132	158	48	B 142	144	A 116	218	A 77	214
D 133	216	52	A 78	213	......	......	A 122	259
D 134	88	37	C 57	178	A 77	214	A 116	218
D 135	188	47	A 78	213	......	......	A 122	259
D 138	177	57	A 78	213	......	......	A 122	259
D 139	103	53	A 78	213	......	......	A 122	259
D 140	164	40	A 78	213	......	......	A 122	259
D 142	125	37	A 78	213	......	......	A 122	259
D 144	133	49	C 68	147	A 78	213	A 116	218
D 146	117	46	C 146	209	A 122	259	A 77	214
D 147	129	34	C 49	194	A 79	219	65	218
D 149	212	45	Yd. 5-6	......	......	......	......	......
D 150	196	54	C 68	147	A 78	213	A 116	218
D 151	139	35	Yd. 6	......	......	......	......	......
D 154	248	54	Yd. 5-6	......	......	......	......	......
D 155	187	45	A 105	200	......	......	A 81	203
D 156	159	48	A 93	201	......	......	A 81	203
D 157	175	42	C 38	213	A 111	204	A 78	213
D 158	101	47	B 159	188	207	169	A 77	214
D 159	165	50	C 40	151	A 77	214	A 116	218
D 160	159	48	A 111	204	......	......	A 77	214
D 162	203	48	C 35	225	A 79	219	65	218
D 164	151	54	B 158	131	226	160	A 77	214
D 165	193	51	A 78	213	......	......	A 122	259
D 166	217	53	C 41	177	A 79	219	65	218
D 167	142	34	C 61	158	A 79	219	65	218
D 168	163	50	C 125	158	A 79	219	A 77	214
D 169	107	31	C 57	178	A 77	214	A 116	218
D 170	175	42	A 93	201	......	......	A 81	203
D 171	224	50	A 111	204	......	......	A 77	214
D 172	221	55	B 252	96	226	160	A 77	214
D 173	121	30	B 154	191	204	182	A 81	203
D 174	174	46	B 158	131	226	160	A 77	214
D 175	102	43	C 46	201	A 77	214	A 116	218
D 177	268	58	A 83	167	......	......	A 122	259
D 178	188	53	A 106	202	......	......	A 122	259
D 180	239	61	C 125	158	A 79	219	A 77	214
D 182	151	42	A 78	213	......	......	A 122	259
D 183	218	52	Yd. 5	......	......	......	......	......
D 184	122	43	C 49	194	A 79	219	65	218
D 186	160	52	H26N	139	204	182	A 81	203
D 187	130	28	Yd. 6	......	......	......	......	......
D 188	179	46	Yd. 7	......	......	......	A 111	204
D 190	176	42	Yd. 7	......	......	......	A 111	204
D 192	217	51	Yd. 7	......	......	......	A 111	204
D 193	193	57	Yd. 7	......	......	......	A 111	204
D 194	177	39	Yd. 7	......	......	......	A 111	204
①......	169.41	46.57	....	183.32	....	205.75	....	224.57
②......	175.59							

① Average of yard. ② Unidentified eggs added.

Table VI gives a record of 33 pullets in yard 8, 1912-13. Their average production was 174.18. They were late hatched, on the average. Their dams' records averaged 196.76, and sire's dams 183.85.

The average of the 108 Barred Rock pullets in yards 6, 7, 8, in 1912-13 was 177.91.

Fifth Generation. The records of the next generation of Barred Rock pullets are shown in Tables VII and VIII, yards B and C. Fifty-

TABLE IX. ANNUAL AND BEST TWO-MONTHS PRODUCTION
Yard L, 1914-15, Barred Plymouth Rocks.

Hen No.	Production 1st year	Best 2 months	Dam	Eggs 1st yr.	D.D.	Eggs 1st yr.	S.D.	Eggs 1st yr.
E 578	172	49	C 50	224	A 77	214	A 122	259
E 580	134	40	Yd. B	......	....	......	A 78	213
E 581	143	43	Yd. B	......	....	......	A 78	213
E 582	188	44	C 76	198	B 154	191	H12N	144
E 591	196	48	Yd. B	......	....	......	A 78	213
E 594	222	46	Yd. B	......	....	......	A 78	213
E 595	158	46	C 35	225	A 79	219	C 69	232
E 596	182	42	Yd. B	......	....	......	A 78	213
E 597	200	47	Yd. B	......	....	......	A 78	213
E 599	187	49	Yd. B	......	....	......	A 78	213
E 602	172	43	C 151	167	A 77	214	A 77	214
E 605	199	50	Yd. B	......	....	......	A 78	213
E 609	195	49	Yd. B	......	....	......	A 78	213
E 611	134	31	C 41	177	A 79	219	Yd. 7	......
E 612	204	59	A 96	204	....	......	A 122	259
E 613	175	49	C 62	175	A 78	213	Yd. 7	......
E 617	221	54	C 50	224	A 77	214	A 122	259
E 618	108	39	Yd. B	......	....	......	A 78	213
E 619	185	50	C 117	186	A 79	219	H12N	144
E 621	189	52	C 50	224	A 77	214	A 122	259
E 622	155	38	C 43	142	A 81	203	A 77	214
E 623	218	48	Yd. B	......	....	......	A 78	213
E 624	189	56	C 168	193	B 151	136	Yd. 7	......
E 625	177	44	C 63	140	A 77	214	Yd. 7	......
E 626	201	43	204	183	....	......	Yd. 7	......
E 627	160	40	C 149	171	A 78	213	Yd. 7	......
E 628	218	43	C 63	140	A 77	214	Yd. 7	......
E 629	186	52	C 41	177	A 79	219	Yd. 7	......
E 630	210	47	C 41	177	A 79	219	Yd. 7	......
E 631	160	37	Yd. B	......	....	......	A 78	213
E 633	114	33	204	183	....	......	Yd. 7	......
E 634	175	41	204	183	....	......	Yd. 7	......
2288	143	42	C 62	175	A 78	213	Yd. 7	......
①	176.36	45.27	....	181.33	....	208.71	....	216.19
②	185.78							

① Average of yard. ② Unidentified eggs added.

two pullets, yard B, averaged 185 eggs; the dams averaged 193.86; dams' dams, 215.05; sires' dams, 220.52. Table VIII shows an average production of 175.59 for the 51 pullets in yard C, with dams' average 183.32. Dams' dams 205.75 and sires' dams, 224.57, high record of 268 eggs, equaled the previous high record for Barred Rocks in the previous year, yard 6. In a few cases the dams' record is low, but with dam's dams high and sire's dam high, they are left in the tabulation.

Sixth Generation. The sixth generation of pullets is shown in Tables IX and X, yards L and P. The former averaged 185.78 and the latter 189.01. The average of dams in yard L was 181.33, of dam's dam 208.71, and sire's dam 216.19. In yard P the average of dams is 213.84, dam's dam 214.81, sire's dam 231.68.

Seventh Generation. A shortage of help and interruptions due to war service were responsible for the small number of pullets available for experimentation for the years 1915-16 and 1916-17. For the year 1915-16, no records can be given for comparison. In 1916-17, yard 18, Table XI, averaged 201.9, with dam's average 247.36, dams' dams 197.44, sires' dams 223.13.

TABLE X. ANNUAL AND BEST TWO-MONTHS PRODUCTION
Yard P. 1914-15, Barred Plymouth Rocks.

Hen No.	Production 1st year	Best 2 months	Dam	1st yr. Eggs	D.D.	Eggs 1st yr.	S.D.	Eggs 1st yr.
E 501	157	47	C 152	195	207	169	A 78	213
E 506	139	38	A 78	213	......	......	C 69	232
E 507	173	44	A 78	213	......	......	C 69	232
E 508	165	49	C 166	233	A 96	204	C 69	232
E 509	136	45	C 152	195	207	169	A 78	213
E 511	135	31	C 112	219	B 154	191	C 64	235
E 512	183	47	C 112	219	B 154	191	C 64	235
E 514	193	38	A 78	213	......	......	C 69	232
E 515	245	52	C 56	210	A 111	204	C 64	235
E 516	197	43	A 78	213	......	......	C 69	232
E 518	152	56	C 166	233	A 96	204	C 69	232
E 519	141	44	C 56	210	A 111	204	C 64	235
E 520	176	45	C 67	187	A 111	204	C 64	235
E 521	246	48	C 146	209	A 122	259	C 69	232
E 523	179	40	C 90	215	B 154	191	C 69	232
E 525	226	54	C 166	233	A 96	204	C 69	232
E 526	140	42	A 77	214	......	......	C 64	235
E 527	155	41	C 146	209	A 122	259	C 69	232
E 529	156	40	A 77	214	......	......	C 64	235
E 531	231	57	C 146	209	A 122	259	C 69	232
E 532	198	48	C 146	209	A 122	259	C 69	232
E 534	222	54	C 146	209	A 122	259	C 69	232
E 535	195	46	C 146	209	A 122	259	C 69	232
E 536	209	51	A 77	214	......	......	C 64	235
E 538	150	42	C 112	219	B 154	191	C 64	235
E 539	182	45	A 78	213	......	......	C 69	232
E 542	238	58	C 146	209	A 122	259	C 69	232
E 543	243	52	C 146	209	A 122	259	C 69	232
E 544	133	45	A 77	214	......	......	C 64	235
E 545	244	49	C 90	215	B 154	191	C 69	232
E 546	218	52	A 78	213	......	......	C 69	232
E 548	173	47	A 77	214	......	......	C 64	235
E 549	176	48	C 56	210	A 111	204	C 64	235
E 550	137	41	C 112	219	B 154	191	C 64	235
E 555	143	41	C 56	210	A 111	204	C 64	235
E 556	153	41	C 56	210	A 111	204	C 64	235
E 557	234	57	C 166	233	A 96	204	C 69	232
E 558	221	56	C 119	241	A 96	204	A 78	213
①	184.06	46.69	....	213 84	....	214.81	....	231.68
②	189.01							

① Average of yard.   ② Unidentified eggs added.

TABLE XI. ANNUAL AND BEST TWO-MONTHS PRODUCTION
Yard 18, 1916-17, Barred Plymouth Rocks.

Hen No.	Production 1st year	Best 2 months	Dam	Eggs 1st yr.	D.D.	Eggs 1st yr.	S.D.	Eggs 1st yr.
G 1	199	52	D 177	268	A 83	167	C 50	224
G 2	203	49	D 92	259	C 35	225	C 50	224
G 3	195	58	E 561	185	......	......	C 50	224
G 4	183	53	E 561	185	....	......	C 50	224
G 5	214	44	D 177	268	A 83	167	C 50	224
G 7	220	50	D 92	259	C 35	225	C 50	224
G 8	195	57	D 92	259	C 35	225	C 50	224
G 9	212	55	D 177	268	A 83	167	C 50	224
G 10	179	48	D 92	259	C 35	225	C 50	224
G 11	185	45	E 543	243	C 146	209	D 52	214
G 12	154	52	D 177	268	A 83	167	C 50	224
①	194.45	51.18	....	247.36	....	197.44	....	223.13
②	201.90							

TABLE XII. ANNUAL AND BEST TWO-MONTHS PRODUCTION
Yard E, 1917-18, Barred Plymouth Rocks.

Hen No.	Production 1st year	Production Best 2 months	Dam	Eggs 1st yr.	D.D.	Eggs 1st yr.	S.D.	Eggs 1st yr.
H 101	201	46	G 7	220	D 92	259	D 52	214
H 102	140	41	Yd. M	......	......	......	D 87	243
H 103	239	51	F 83	179	C 56	210	E 543	243
H 104	205	44	G 1	199	D 177	268	D 52	214
H 105	204	50	G 7	220	D 92	259	D 52	214
H 106	227	49	F 90	209	D 154	248	E 543	243
H 107	211	45	G 1	199	D 177	208	D 52	214
H 108	158	44	F 84	169	D 156	159	E 543	243
H 109	133	43	G 10	179	D 92	259	D 52	214
H 111	214	52	G 3	195	E 561	185	D 52	214
H 112	204	41	G 11	185	E 543	243	D 52	214
H 113	182	47	F 90	209	D 154	248	E 543	243
H 114	206	54	Yd. M	......	......	......	D 87	243
H 115	74	31	F 18	271	A 06	204	D 177	268
H 116	208	48	Yd. M	......	......	......	D 87	243
H 118	94	28	Yd. M	......	......	......	D 87	243
H 119	181	46	Yd. N	......	......	......	D 177	268
H 120	211	44	Yd. N	......	......	......	D 177	268
H 121	185	45	Yd L	......	......	......	D 39	270
H 122	240	48	F 4	192	D 100	213	E 543	243
H 123	189	54	Yd. L	......	......	......	D 39	270
H 124	207	53	Yd. M	......	......	......	D 87	243
H 125	113	33	Yd. M	......	......	......	D 87	243
H 126	184	52	Yd. M	......	......	......	D 87	243
H 127	162	44	G 166	243	D 136	223	D 87	243
H 128	217	46	Yd. N	......	......	......	D 177	268
H 129	54	27	Yd. N	......	......	......	D 177	268
H 130	251	52	Yd. N	......	......	......	D 177	268
H 131	191	50	Yd. N	......	......	......	D 177	268
H 132	242	47	F 9	161	C 112	219	D 84	236
H 133	277	57	Yd. N	......	......	......	D 177	268
H 134	269	58	Yd. N	......	......	......	D 177	268
H 135	91	32	Yd. N	......	......	......	D 177	268
H 136	219	41	F 4	192	D 100	213	E 543	243
H 137	224	44	F 89	182	D 136	223	E 543	243
H 138	234	49	Yd. N	......	......	......	D 177	268
H 139	158	46	Yd. M	......	......	......	D 87	243
H 140	210	42	Yd. M	......	......	......	D 87	243
H 141	135	35	Yd. M	......	......	......	D 87	243
H 142	268	53	Yd. N	......	......	......	D 177	268
H 144	210	51	Yd. 0	......	......	......	E 561	185
H 145	270	56	Yd. N	......	......	......	D 177	268
H 146	232	54	Yd. 0	......	......	......	E 561	185
H 147	260	47	Yd. N	......	......	......	D 177	268
H 148	191	50	Yd. 0	......	......	......	E 561	185
H 150	183	45	Yd. 0	......	......	......	E 561	185
H 151	153	44	Yd. 0	......	......	......	E 561	185
H 153	224	52	Yd. N	......	......	......	D 177	268
H 154	251	53	Yd. N	......	......	......	D 177	268
H 155	244	54	Yd. N	......	......	......	D 177	268
H 156	226	53	Yd. N	......	......	......	E 561	185
H 157	225	50	Yd. 0	......	......	......	E 561	185
H 159	220	50	Yd. N	......	......	......	D 177	268
H 160	216	46	Yd. N	......	......	......	D 177	268
H 161	211	47	Yd. 0	......	......	......	E 561	185
H 162	208	60	Yd. N	......	......	......	D 177	268
H 163	215	49	Yd. M	......	......	......	D 87	243
H 165	229	52	Yd. L	......	......	......	D 39	270
H 166	282	54	Yd. L	......	......	......	D 39	270
H 167	244	54	Yd. L	......	......	......	D 39	270
H 168	200	47	Yd. L	......	......	......	D 39	270
H 169	203	46	Yd. L	......	......	......	D 39	270
①	202.40	47.19	......	200.23	......	229.47	......	244.58
②	214.63							

① Average of yard. ② Unidentified eggs added.

Eighth Generation. In the eighth generation, Table XII, there are 62 pullets in yard E with average records of 214.63 eggs. This shows a rather remarkable increase over previous year, possibly more than would reasonably be expected in one generation as an average. Environmental conditions were more favorable than in some of the previous years. The dams averaged 200.23, dams' dams 229.47, sires' dams 244.58. In several cases the dam's record is not known, only the yard in which the dam was bred. In those cases, however, the record of the dam was high, for all the hens in the yard were of good records.

Effect of Close Confinement. To determine how much of a factor methods of housing and yarding are in our breeding experiments, a pen of twenty Barred Plymouth Rock pullets from our experimental flock were confined throughout the year, 1915-16, in a house 8 by 12 feet in size. It was the same style of house in which our experimental

Fig. 2. A pen of the Oregon Station's strain of Barred Plymouth Rocks. This pen tied for second place with a pen of the Station's "Oregons" in the egg-laying contest at the Panama-Pacific International Exposition, San Francisco. The Station sent three pens and won the first three prizes.

flocks were kept, but the experimental flocks had the liberty of outdoor yards at all seasons. This pen averaged 182.6 eggs a year with high record of 289 and low of 85. There were no deaths during the year. This record is practically the same as for the experimental flocks of the same breeding, and indicates that conditions as to housing and yarding are not important factors, so far as indoor and outdoor management are concerned.

It is understood, of course, that this test and these conclusions have nothing to do with the question of management as it affects breeding stock or reproduction. Because of radical difference in environment, the record of this pen is not included in tabulation in comparing the effect of breeding.

RESULTS WITH WHITE LEGHORNS

Foundation Stock. The production of the original stock of White Leghorns is given in Table XIII. These pullets were bought from different breeders in Oregon, and fifty were used as foundation stock. The production records for the term of the experiment embraced 386 pullets.

In the first year, the average production was 106.88 eggs a hen. The highest record was 183 and the lowest 2 eggs.

Second Generation. In the second year (Table XIV) there were only 21 pullets with records, the highest being 229 eggs and the lowest 3, average 104.67 eggs. In this case, no "floor" eggs are counted, there being no record of unidentified eggs. Were these eggs accounted for the record of these pullets would doubtless have equalled the record of the previous year. These pullets were not pedigreed birds but produced promiscuously from the original flock and from males of unknown ancestry. Some of these pullets were later used for breeding. Hen O34 was our first 200-egg Leghorn, and was the foundation of our high-producing Leghorn stock.

Fig. 3. Pen 3 of White Leghorns, 1914-15; record 230.12. See Table XVII.

Third Generation. In the third year, 1910-11 (Table XV), there are only 10 pullets available. These were not pedigreed, but hatched from the foundation stock after the poor layers had been culled out. They averaged 164.10 eggs with highest 240 and lowest 98, hen A27 making a record of 240. This hen lived until she was 8 years and one month old, and in that time had laid in trap-nests, 1188 eggs. All of the present Station Leghorns trace pedigrees back to this hen.

Fourth Generation. Table XVI gives the record of fourteen White Leghorn pullets included in yard 9 of Oregons, 1912-13. They averaged 208.93 eggs with dam's average 223.85 and sire's dam 229. Three of these birds it will be noticed were from an inbred mating of dam to cockerel. Their numbers are C526, C551, and C584.

Fifth Generation. In 1914-15 (Table XVII) sixteen pedigreed pullets in yard 3 averaged 230.12 eggs, with dams' record of 242.19, dams' dams 222.33, and sires' dams 233.13. The highest record was 302 and the lowest 148. This yard produced our first 300-egg Leghorn. In the

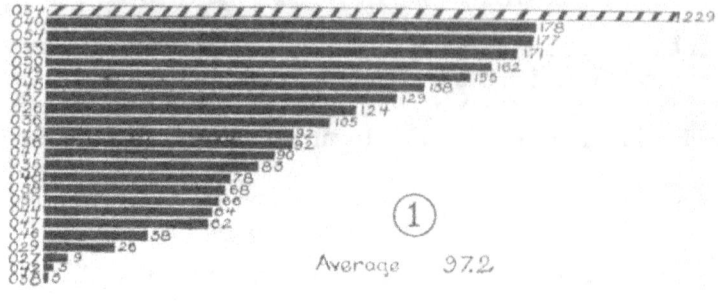

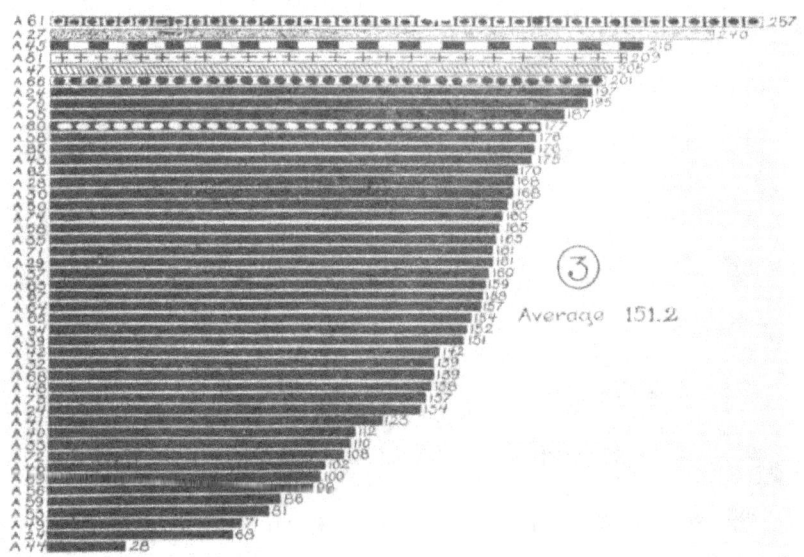

Fig. 5. These charts show graphically a progressive increase in production by selecting the best layers in the flock for breeding. Chart 1 is the record of an unselected pen of White Leghorns; Chart 2, an unselected pen of crosses; Chart 3, a partly selected pen of White Leghorns and crosses; and Chart 4 a pen bred from high producers or pedigreed stock. Chart 4 is the result of breeding from the best hens represented in Charts 1, 2, and 3, all better than 200-egg hens except one, and this one laid 234 in second year. Full sisters in Chart 4 are represented by lines of similar pattern. A son of O34, Chart 1, was the sire of all the pullets except daughters of 250, this hen being inbred to her son. There are also three daughters of O34 inbred to son. Fewer poor layers and more good layers are shown in Chart 4 than in the flocks from which their parents were selected, but the increased production has not been followed by uniformity in production. The average production of the daughters does not equal the production of the parents, but there are individual pullet records higher as well as lower than the production of the parents.

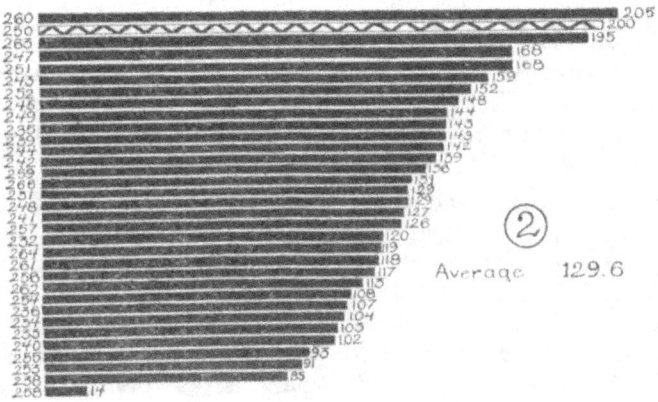

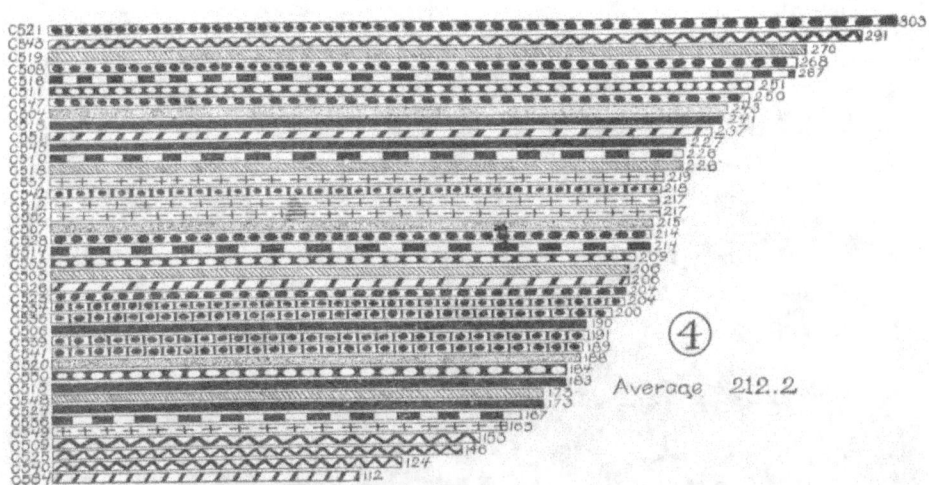

Fig. 5. These charts show graphically a progressive increase in production by selecting the best layers in the flock for breeding. Chart 1 is the record of an unselected pen of White Leghorns; Chart 2, an unselected pen of crosses; Chart 3, a partly selected pen of White Leghorns and crosses; and Chart 4 a pen bred from high producers or pedigreed stock. Chart 4 is the result of breeding from the best hens represented in Charts 1, 2, and 3, all better than 200-egg hens except one, and this one laid 234 in second year. Full sisters in Chart 4 are represented by lines of similar pattern. A son of O34, Chart 1, was the sire of all the pullets except daughters of 250, this hen being inbred to her son. There are also three daughters of O34 inbred to son. Fewer poor layers and more good layers are shown in Chart 4 than in the flocks from which their parents were selected, but the increased production has not been followed by uniformity in production. The average production of the daughters does not equal the production of the parents, but there are individual pullet records higher as well as lower than the production of the parents.

same year 42 pullets in yard O (Table XVIII) averaged 216.72 with dams' record 236.97, dams' dams 221.68, sires' dams 237.25.

Sixth Generation. In 1916-17, yard P (Table XIX), 49 pullets average 188.96 with dams' record at 209.74, dams' dams 247.27, sires' dams 251.71 eggs. The large proportion of these pullets were early hatched, and some of them went through the moult in the fall, which doubtless reduced the production.

In yard Q (Table XX) the same year 57 pullets averaged 172.65, dams 211.75, dams' dams 241.56, sires' dams 258.74. This yard was not under favorable conditions for high record. There was not much uniformity in the ages; some were early hatched and moulted, and others were late and did not start to lay early enough to make a good record. In the same year 11 pullets of yard 15 (Table XXI) averaged 225.27, with dams 227.64, dams' dams 245.09, and sires' dams 242.18. These 11 pullets were of practically the same production pedigree as P and Q. The fact that they did not moult probably explains their high record.

Fig. 5. A pen of the Oregon Station's pedigreed strain of White Leghorns. This pen won first place in a year's egg-laying contest at the Panama-Pacific International Exposition, San Francisco.

Seventh Generation. In 1917-18, sixty pullets in yard F (Table XXII) averaged 222.22 eggs, with high records of 300 and low of 23. Unfortunately there is only a yard pedigree for the dam for most of these hens. Yard R was made up of high-producing stock that averaged 214.14 eggs. The 19 hens with individual pedigrees had dams of averages of 235.42 eggs. The sire's dam, however, in all cases was from our highest record Leghorn, E248, 302 eggs in first year. See Table XVII. The production of yard F would, no doubt, have been higher but for an accident or oversight in May when the fowls had no water for two days. This caused a reduction from about sixty percent to

TABLE XIII.   ANNUAL AND BEST TWO-MONTHS PRODUCTION
Yards 4 and 5, 1908-9, White Leghorns.

Hen No.	Production 1st year	Production Best 2 months	Hen No.	Production 1st year	Production Best 2 months	Hen No	Production 1st year	Production Best 2 months
1	177	50	26	112	37	43	97	27
2	83	34	27	150	40	45	119	33
4	38	13	28	14	5	46	111	35
5	127	38	29	72	25	49	24	12
6	87	26	30	116	28	50	104	33
8	170	42	31	90	29	51	48	17
9	95	33	32	136	49	52	164	41
10	140	42	33	119	33	54	91	28
11	146	42	34	65	22	55	15	13
13	114	35	35	166	43	56	169	45
15	139	44	36	87	36	58	74	37
18	143	40	37	66	30	()	158	41
19	107	37	38	155	38	60	132	34
21	154	38	39	128	40	61	71	35
22	76	29	40	183	45	62	34	14
23	110	33	41	2	2	64	46	36
25	131	42	42	122	35			

Average of yard		106.14	32.54
Unidentified eggs added		106.88	....

TABLE XIV.   ANNUAL AND BEST TWO-MONTHS PRODUCTION
Yard 9, 1909-10, White Leghorns.

Hen No.	Production 1st year	Production Best 2 months	Hen No.	Production 1st year	Production Best 2 months	Hen No.	Production 1st year	Production Best 2 months
O 33	171	42	O 41	90	40	O 48	78	31
O 34	229	47	O 42	5	2	O 49	155	36
O 35	83	23	O 43	92	22	O 50	162	38
O 36	105	41	O 44	64	34	O 54	178	44
O 37	129	29	O 45	150	37	O 56	92	31
O 38	3	2	O 46	38	19	O 57	66	33
O 40	178	48	O 47	62	22	O 58	68	24

Average of yard		104.67	30.71

TABLE XV.   ANNUAL AND BEST TWO-MONTHS PRODUCTION
Yard 1, 1910-11, White Leghorns.

Hen No.	Production 1st year	Production Best 2 months	Hen No.	Production 1st year	Production Best 2 months	Hen No.	Production 1st year	Production Best 2 months
A 27	240	51	A 45	215	47	A 56	98	34
A 38	176	47	A 50	167	42	A 67	158	42
A 41	123	26	A 51	209	48	A 69	118	38
A 42	142	38						

Average of yard		164.6	41.3

TABLE XVI. ANNUAL AND BEST TWO-MONTHS PRODUCTION
Yard 9, 1912-13, White Leghorns.

Hen No.	Production 1st year	Best 2 months	Dam	Eggs 1st year	D.D.	Eggs 1st yr.	S.D.	Eggs 1st year
C 504	243	51	A 27	240	....	....	O 34	229
C 507	215	53	A 27	240	....	....	O 34	229
C 510	226	47	A 45	215	....	....	O 34	229
C 514	214	52	A 45	215	....	....	O 34	229
C 515	241	53	A 27	240	....	....	O 34	229
C 516	267	54	A 45	215	....	....	O 34	229
C 520	188	47	A 27	240	....	....	O 34	229
C 526	206	47	O 34	229	....	....	O 34	229
C 536	167	44	A 45	215	....	....	O 34	229
C 537	219	48	A 51	209	....	....	O 34	229
C 549	163	36	A 51	209	....	....	O 34	229
C 551	225	51	O 34	229	....	....	O 34	229
C 552	217	44	A 51	209	....	....	O 34	229
C 584	119	40	O 34	229	....	....	O 34	229
①	207.85	47.64	......	223.85	....	....	......	229.00
②	208.93							

① Average of yard.   ② Unidentified eggs added.

TABLE XVII. ANNUAL AND BEST TWO-MONTHS PRODUCTION
Yard 3, 1914-15, White Leghorns.

Hen No.	Production 1st year	Best 2 months	Dam	Eggs 1st yr.	D.D.	Eggs 1st yr.	S.D.	Eggs 1st yr.
E 247	271	57	C 516	267	A 45	215	A 27	240
E 248	302	58	C 516	267	A 45	215	A 27	240
E 249	209	49	A 27	240	....	......	O 34	229
E 250	213	48	C 504	243	A 27	240	O 34	229
E 251	237	50	A 27	240	....	......	O 34	229
E 252	229	53	O 34	229	....	......	A 27	240
E 253	233	52	A 27	240	....	......	O 34	229
E 254	205	50	C 515	241	A 27	240	O 34	229
E 255	276	60	C 516	267	A 45	215	A 27	240
E 256	195	50	A 27	240	....	......	O 34	229
E 257	212	47	C 504	243	A 27	240	O 34	229
E 258	234	47	O 34	229	....	......	A 27	240
E 259	189	50	C 496	195	B 228	167	O 34	229
E 260	249	52	A 27	240	....	......	O 34	229
E 261	230	49	B 12	251	O 34	229	A 27	240
E 262	148	47	C 504	243	A 27	240	O 34	229
①	227.00	51.19	......	242.19	...	222.33	....	233.13
②	230.12							

① Average of yard.   ② Unidentified eggs added.

forty percent in a few days, and it was a week or more before they got back to former production.

In the same year, yard G (Table XXIII), 56 pullets averaged 201.85, and their dams averaged 234.5, dams' dams 247.95, sires' dams 236.81. Highest record 298, lowest, 56. These pullets were later-hatched and did not come to laying at as favorable a date as yard F. Had the conditions been favorable, the production of this yard would, no doubt, have been higher.

As it is, the increase in production in the seven generations of White Leghorns was 105.51, the production of the last year lacking just one egg of being double the production of the first generation or foundation stock.

TABLE XVIII. ANNUAL AND BEST TWO-MONTHS PRODUCTION
Yard O, 1914-15, White Leghorns.

Hen No.	Production 1st year	Production Best 2 months	Dam	Eggs 1st yr.	D.D.	Eggs 1st yr.	S.D.	Eggs 1st yr.
E 1	270	51	C 516	267	A 45	215	A 27	240
E 2	219	52	Yd. F	*	....	......	A 27	240
E 3	248	54	A 27	240		......	O 34	229
E 5	230	52	C 516	267	A 45	215	A 27	240
E 6	177	50	C 551	214	O 34	229	A 27	210
E 7	270	54	B 12	251	O 34	229	A 27	240
E 8	215	49	C 516	267	A 45	215	A 27	240
E 9	259	51	B 12	251	O 34	229	A 27	240
E 12	244	49	Yd. F	*	....	......	A 27	240
E 13	258	48	B 12	251	O 34	229	A 27	240
E 14	247	56	C 516	267	A 45	215	A 27	240
E 15	232	50	C 504	243	A 27	240	O 34	229
E 16	203	49	Yd. F	*	....	......	A 27	240
E 17	211	43	Yd. F	*	....	......	A 27	240
E 18	229	43	A 27	240	....	......	O 34	229
E 19	235	50	B 12	251	O 34	229	A 27	240
E 20	243	47	C 516	267	A 45	215	A 27	240
E 21	259	48	B 12	251	O 34	229	A 27	240
E 24	182	47	Yd. F	*	....	......	A 27	240
E 25	211	50	B 12	251	O 34	229	A 27	240
E 26	219	44	C 504	243	A 27	240	O 34	229
E 27	199	52	B 12	251	O 34	229	A 27	240
E 28	184	45	Yd. F	*	....	......	A 27	240
E 29	253	55	Yd. F	*	....	......	A 27	240
E 30	108	43	C 464	175	B 228	167	O 34	229
E 32	185	50	C 516	267	A 45	215	A 27	240
E 33	270	56	C 504	243	A 27	240	O 34	229
E 34	198	52	C 515	241	A 27	240	O 34	229
E 35	100	35	C 516	267	A 45	215	A 27	240
E 36	164	46	D 718	254	C 515	241	......	......
E 37	217	54	D 399	144	B 9	180	....	......
E 38	128	44	D 327	162	B 3	224	......	......
E 40	245	48	B 12	251	O 34	229	A 27	240
E 41	163	47	D 327	162	B 3	224	......	......
E 42	239	54	A 27	240	....	......	O 34	229
E 43	249	51	D 718	254	C 515	241	......	......
E 44	229	51	B 12	251	O 34	229	A 27	240
E 46	169	49	A 51	209	......	......	A 27	240
E 47	156	46	C 516	267	A 45	215	A 27	240
E 49	219	49	C 464	175	B 228	167	O 34	229
E 50	248	52	C 516	267	A 45	215	A 27	240
E 53	156	42	D 731	193	C 504	243	....	......
①	212.86	49.24	....	236.97	....	221.68	....	237.25
②	216.72							

① Average of yard. ② Unidentified eggs added.
* Pen of high producing birds laying from 101 to 276.

TABLE XIX. ANNUAL AND BEST TWO-MONTHS PRODUCTION
Yard P, 1916-17, White Leghorns.

Hen No.	Production 1st year	Production Best 2 months	Dam	Eggs 1st yr.	D.D.	Eggs 1st yr.	S.D.	Eggs 1st yr.
G 620	179	36	E 253	233	A 27	240	D 343	220
G 621	255	48	F 595	73(D)	D 718	254	A 27	240
G 622	155	51	E 253	233	A 27	240	D 343	220
G 623	239	50	F 592	211(D)	D 718	254	A 27	240
G 624	183	48	E 257	212	C 504	243	D 343	220
G 626	177	45	E 257	212	C 504	243	D 343	220
G 627	146	51	F 594	176	D 718	254	A 27	240
G 628	191	49	F 594	176	D 718	254	A 27	240
G 630	202	46	E 253	233	A 27	240	D 343	220
G 631	183	50	F 622	248	Yd. 0	......	A 27	240
G 632	222	52	E 254	205	C 515	241	D 343	220
G 633	167	48	F 643	208(D)	D 240	245	A 27	240
G 634	158	50	Yd. 15	*	......	......	O 34	229
G 635	197	55	F 631	197	Yd. 0	......	A 27	240
G 637	160	41	Yd. 15	*	......	......	O 34	229
G 638	208	40	Yd. 15	*	......	......	O 34	229
G 639	225	47	Yd. 15	*	......	......	O 34	229
G 640	211	48	Yd. K	**	......	......	A 27	240
G 641	209	49	Yd. I	†	......	......	A 27	240
G 642	176	41	Yd. 15	*	......	......	O 34	229
G 643	220	54	Yd. K	**	......	......	A 27	240
G 644	243	53	Yd. 16	‡	......	......	C 515	241
G 646	172	48	Yd. K	**	......	......	A 27	240
G 648	129	41	Yd. K	**	......	......	A 27	240
G 652	184	44	E 255	276	C 516	267	D 343	220
G 658	165	44	Yd. I	†	......	......	A 27	240
G 659	197	52	Yd. I	†	......	......	A 27	240
G 661	197	47	Yd. I	†	......	......	A 27	240
G 662	184	48	Yd. K	**	......	......	A 27	240
G 663	197	46	H7ON	161	A 27	240	E 248	302
G 664	185	50	H7ON	161	A 27	240	E 248	302
G 666	132	42	E 255	276	C 516	267	D 343	220
G 667	234	47	E 260	249	A 27	240	D 343	220
G 668	264	51	A 27	240			E 248	302
G 671	139	49	H15P	147	A 27	240	E 248	302
G 672	227	52	D 718	254	C 515	241	A 27	240
G 673	163	49	E 260	199	D 718	254	E 248	302
G 674	244	51	A 27	240			E 248	302
G 675	156	49	C 590	193	A 27	240	E 248	302
G 676	212	52	C 590	193	A 27	240	E 248	302
G 677	148	47	C 590	193	A 27	240	E 248	302
G 678	129	40	C 590	193	A 27	240	E 248	302
G 679	215	54	H9P	162	A 27	240	E 248	302
G 680	143	40	C 590	193	A 27	240	E 248	302
G 681	137	40	H83N	139	A 27	240	E 248	302
G 682	215	50	E 248	302	C 516	267	D 343	220
G 683	211	51	E 248	302	C 516	267	D 343	220
G 684	127	47	E 248	302	C 516	267	D 343	220
G 685	178	50	H83N	139	A 27	240	E 248	302
①	187.55	47.61	......	209.74	......	247.27	......	251.71
②	188.96							

① Average of yard. ② Unidentified eggs added.
  * Pen of high producing birds laying from 173 to 267.
 ** Pen of high producing birds laying from 106 to 271.
  † Pen of high producing birds laying from 167 to 270.
  ‡ Pen of high producing birds which were at Panama-Pacific Exposition Contest during pullet year and laid from 120 to 204 eggs. Most of these birds laid more in the second year, the highest being 242.

TABLE XX. ANNUAL AND BEST TWO-MONTHS PRODUCTION
Yard Q, 1916-17, White Leghorns.

Hen No.	Production 1st year	Best 2 months	Dam	Eggs 1st yr.	D.D.	Eggs 1st yr.	S.D.	Eggs 1st yr.
G 720	169	42	F 589	170	C 589	211	A 27	240
G 721	192	49	F 625	228	Yd.0	.....	A 27	240
G 722	204	42	E 250	213	C 504	243	D 343	220
G 724	189	47	F 665	205	C 589	211	A 27	240
G 725	231	43	F 655	231	D 268	146	A 27	240
G 726	204	45	F 649	159	C 590	193	A 27	240
G 727	141	39	E 230	199	D 718	254	E 248	302
G 730	188	46	Yd 12	*	....	......	A 27	240
G 721	204	42	F 750	17	C 590	193	D 718	254
G 732	237	49	Yd I	**	....	.....	A 27	240
G 735	201	47	Yd I	**	....	......	A 27	240
G 736	207	49	Yd I	**	....	......	A 27	240
G 738	212	57	Yd K	†	....	......	A 27	240
G 739	181	42	Yd I	**	....	.....	A 27	240
G 741	114	45	H70N	161	A 27	240	E 248	302
G 742	130	48	H70N	161	A 27	240	E 248	302
G 743	114	36	F 596	209	....	......	O 34	229
G 744	206	50	F 719	243	E 257	212	C 589	170+
G 745	163	18	F 607	208	....	......	O 34	229
G 746	144	55	E 230	199	D 718	254	E 248	302
G 747	179	46	E 230	199	D 718	254	E 248	302
G 748	173	48	E 214	135	O 34	229	E 248	302
G 749	196	47	E 255	276	C 516	267	D 343	220
G 750	130	41	E 269	254	Yd.F Leg.	.....	E 248	302
G 751	121	35	A 27	240	....	.....	E 248	302
G 752	136	34	E 21	259	B 12	251	A 27	240
G 753	166	51	E 3	248	A 27	240	A 27	240
G 754	103	46	H83N	139	A 27	240	E 248	302
G 755	126	42	H15P	147	A 27	240	E 248	302
G 756	130	46	H83N	139	A 27	240	E 248	302
G 757	127	37	H15P	147	A 27	240	E 248	302
G 758	190	50	E 255	276	C 516	267	D 343	220
G 760	143	45	F 984	194	....	.....	A 27	240
G 761	174	47	E 255	276	C 516	267	D 343	220
G 762	111	33	H15P	147	A 27	240	E 248	302
G 763	177	49	H9P	162	A 27	240	E 248	302
G 764	196	53	E 13	258	B 12	251	A 27	240
G 765	241	47	E 247	271	C 516	267	D 343	220
G 767	256	51	E 247	271	C 516	267	D 343	220
G 768	187	46	E 260	249	A 27	240	D 343	220
G 769	103	36	E 230	199	D 718	254	E 248	302
G 770	202	58	E 260	249	A 27	240	D 343	220
G 771	67	32	H83N	139	A 27	240	E 248	302
G 772	217	49	D 718	254	C 515	241	A 27	240
G 773	185	45	E 248	302	C 516	267	D 343	220
G 774	186	40	E 248	302	C 516	267	D 343	220
G 775	140	45	A 27	240	....	.....	E 248	302
G 776	219	48	E 247	271	C 516	267	D 343	220
G 777	188	41	E 247	271	C 516	267	D 343	220
G 778	171	44	E 260	249	A 27	240	D 343	220
G 779	63	34	H83N	139	A 27	240	E 248	302
G 781	228	52	A 27	240	....	.....	E 248	302
G 782	167	44	H15P	147	A 27	240	E 248	302
G 783	121	39	H15P	147	A 27	240	E 248	302
G 784	217	46	H83N	139	A 27	240	E 248	302
G 785	219	52	E 248	302	C 516	267	D 343	220
G 786	95	27	H83N	139	A 27	240	E 248	302
①	169.84	44.77	....	211.75	....	241.56	....	258.74
②	172.65							

① Average of yard. ② Unidentified eggs added.
   * Pen of high producing birds laying from 130 to 263 eggs.
  ** Pen of high producing birds laying from 167 to 270 eggs.
   † Pen of high producing birds laying from 106 to 271 eggs.

TABLE XXI.   ANNUAL AND BEST TWO-MONTHS PRODUCTION
Yard 15, 1916-17, White Leghorns.

Hen No.	Production 1st year	Best 2 months	Dam	Eggs 1st yr.	D.D.	Eggs 1st yr.	S.D.	Eggs 1st yr.
G 501	245	58	F 255	276	C 516	267	D 543	220
G 502	215	52	E 247	271	C 516	267	D 543	220
G 503	234	55	E 248	302	C 516	267	D 343	220
G 506	210	55	E 255	276	C 516	267	D 343	220
G 507	184	47	E 258	234	O 34	229	D 343	220
G 508	228	48	E 9	259	B 12	251	A 27	240
G 509	196	52	F 701	172	C 551	214	A 27	240
G 510	160	42	F 701	172	C 551	214	A 27	240
G 512	238	52	E 3	248	A 27	240	A 27	240
G 513	234	51	H15P	147	A 27	240	E 248	302
G 514	221	56	H15P	147	A 27	240	E 248	302
①	215.00	51.63	....	227.64	......	245.09	....	242.13
②	225.27							

① Average of yard.   ② Unidentified eggs added.

TABLE XXII. ANNUAL AND BEST TWO-MONTHS PRODUCTION
Yard F, 1917-18, White Leghorns.

Hen No.	Production 1st year	Best 2 months	Dam	Eggs 1st yr.	D.D.	Eggs 1st yr.	S.D.	Eggs 1st yr.
H 502	234	48	D 343	220	B 2	223	E 248	302
H 503	208	52	Yd. R				E 248	302
H 504	239	45	Yd. R				E 248	302
H 505	195	44	Yd. R				E 248	302
H 507	230	49	Yd. R				E 248	302
H 508	240	51	Yd. R				E 248	302
H 509	167	36	Yd. R				E 248	302
H 510	230	51	Yd. R				E 243	302
H 511	265	54	Yd. R				E 248	302
H 512	243	50	F 607	208			E 248	302
H 513	248	49	D 343	220	B 2	223	E 248	302
H 514	167	38	Yd. R				E 248	302
H 515	272	57	Yd. R				E 248	302
H 516	183	52	Yd. R				E 248	302
H 517	238	49	Yd. R				E 248	302
H 518	201	49	Yd. R				E 248	302
H 519	238	50	Yd. R				E 248	302
H 521	256	51	Yd. R				E 248	302
H 522	271	50	Yd. R				E 248	302
H 523	241	50	Yd. R				E 248	302
H 524	254	52	Yd. R				E 248	302
H 525	168	48	Yd. R				E 248	302
H 526	217	50	Yd. R				E 248	302
H 527	213	49	Yd. R				E 248	302
H 528	244	47	Yd. R				E 248	302
H 529	217	50	Yd. R				E 248	302
H 530	235	50	Yd. R				E 248	302
H 531	119	37	E 9	259	B 12	251	E 248	302
H 532	240	51	Yd. R				E 248	302
H 533	298	57	Yd. R				E 248	302
H 534	145	44	E 21	259	B 12	251	E 248	302
H 535	218	51	Yd. R				E 248	302
H 536	256	49	Yd. R				E 248	302
H 537	177	44	Yd. R				E 248	302
H 538	200	43	F 906	170	C 516	267	E 248	302
H 539	208	47	Yd. R				E 248	302
H 540	251	59	Yd. R				E 248	302
H 541	163	47	Yd. R				E 248	302
H 542	149	40	Yd. R				E 248	302
H 543	247	53	Yd. R				E 248	302
H 544	279	51	E 9	259	B 12	251	E 248	302
H 545	300	59	Yd. R				E 248	302
H 548	239	50	Yd. R				E 248	302
H 549	204	42	Yd. R				E 248	302
H 550	178	43	Yd. R				E 248	302
H 551	252	51	E 21	259	B 12	251	E 248	302
H 552	190	44	Yd. R				E 248	302
H 553	227	50	F 609	235	A 27	240	E 248	302
H 555	154	49	Yd. R				E 248	302
H 556	23	9	Yd. R				E 248	302
H 557	265	50	D 343	220	B 2	225	E 248	302
H 558	234	47	E 21	259	B 12	251	E 248	302
H 559	260	48	D 343	220	B 2	223	E 248	302
H 560	246	49	D 343	220	B 2	223	E 248	302
H 561	191	43	D 343	220	B 2	223	E 248	302
H 562	237	49	E 21	259	B 12	251	E 248	302
H 563	187	47	E 9	259	B 12	251	E 248	302
H 564	243	49	E 21	259	B 12	251	E 248	302
H 565	175	46	D 343	220	B 2	223	E 248	302
H 566	224	48	E 3	248	A 27	240	E 248	302
①	218.21	47.78	....	235.42	....	239.78	....	302.00
②	222.22							

① Average of yard.   ② Unidentified eggs added.

TABLE XXIII. ANNUAL AND BEST TWO-MONTHS PRODUCTION
Yard G, 1917-18, White Leghorns.

Hen No.	Production 1st year	Production Best 2 months	Dam	Eggs 1st yr.	D.D.	Eggs 1st yr.	S.D.	Eggs 1st yr.
H 601	249	53	A 27	240	....	......	E248	302
H 605	184	45	Yd. P	......	....	......	A 27	240
H 607	173	53	Yd. P	......	....	......	A 27	240
H 610	222	46	......	......	....	......	......	......
H 612	211	55	G 501	245	E 255	276	C590	193
H 613	226	52	Yd. Q	......	....	......	A 27	240
H 614	173	41	Yd 15	......	....	......	C590	193
H 616	236	47	Yd. Q	......	....	......	A 27	240
H 619	211	49	Yd. P	......	....	......	A 27	240
H 620	223	55	Yd. 8	......	....	......	B 14	215
H 621	219	56	Yd. Q	......	....	......	A 27	240
H 622	117	37	Yd. V1	......	....	......	E230	199
H 623	221	51	Yd. P	......	....	......	A 27	240
H 624	184	47	Yd. Q	......	....	......	A 27	240
H 626	210	53	Yd. P	......	....	......	A 27	240
H 627	252	52	Yd. Q	......	....	......	A 27	240
H 628	218	49	Yd. U	......	....	......	C189	135
H 629	254	54	Yd. Q	......	....	......	A 27	240
H 631	157	39	Yd. Q	......	....	......	A 27	240
H 632	204	52	......	......	....	......	......	......
H 633	216	50	G 682	215	E 248	302	A 27	240
H 635	154	53	Yd. V1	......	....	......	E230	199
H 636	272	54	Yd. Q	......	....	......	A 27	240
H 637	215	52	G 530	226	H 9P	162	D718	254
H 639	197	49	Yd. Q	......	....	......	A 27	240
H 640	247	50	Yd. Q	......	....	......	A 27	240
H 641	200	50	Yd. Q	......	....	......	A 27	240
H 642	155	54	Yd. 6	......	....	......	A 27	240
H 643	204	50	Yd. V1	......	....	......	E230	199
H 644	168	46	E 249	209	A 27	240	A 27	240
H 645	211	41	Yd. V1	......	....	......	E230	199
H 646	195	54	Yd. V2	......	....	......	C551	214
H 648	217	50	E 248	302	C 516	267	A 27	240
H 649	112	43	A 27	240	....	......	E248	302
H 650	182	52	A 27	240	....	......	E248	302
H 651	137	44	E 248	302	C 516	267	A 27	240
H 652	195	46	F 609	235	A 27	240	E248	302
H 653	267	53	Yd. U	......	....	......	C189	135
H 654	204	44	Yd. U	......	....	......	C189	135
H 655	250	50	Yd. U	......	....	......	C189	135
H 656	199	43	Yd. U	......	....	......	C189	135
H 657	224	47	Yd. U	......	....	......	C189	135
H 658	210	44	E 21	259	B 12	251	E248	302
H 660	212	48	E 21	259	B 12	251	E248	302
H 662	298	59	G 528	272	E 247	271	D718	254
H 663	161	51	D 343	220	B 2	223	E248	302
H 664	181	44	F 596	209	....	......	E248	302
H 665	239	49	D 343	220	B 2	223	E248	302
H 666	172	46	G 529	231	E 260	249	D718	254
H 667	252	53	G 529	231	E 260	249	D718	254
H 668	169	45	E 317	243	A 27	240	E248	302
H 669	156	51	G 529	231	E 260	249	D718	254
H 672	139	42	G 743	114	F 596	209	A 27	240
H 673	183	48	E 317	243	A 27	240	E248	302
H 674	186	46	F 607	208	....	.....	E248	302
H 675	56	33	G 503	234	E 248	302	C590	193
①	199.63	48.57	....	234.50	....	247.95	....	236.81
②	201.85							

① Average of yard.　② Unidentified eggs added.

CROSSING AND INHERITANCE

Foundation Stock. In 1908-09 reciprocal crosses were made with White Leghorns and Barred Plymouth Rocks, the same sires being used as were used in producing the White Leghorn pullets of yard 9, and the Barred Rocks of yard 6, of the second year. That is to say, in the original yards 4 and 5 there were both Leghorn pullets and Barred Rock pullets, but in one yard there were Leghorn males, and in the other Barred Rock males. Any difference, therefore, in production of the pullets from these matings whether Leghorns or Barred Rocks or crosses, can not be due to differences in males.

Second Generation. In the second generation or crosses there were two yards of pullets. Yard 7 (Table XXIV) averaged 131.06 and yard 8 (Table XXV) 140.36 eggs. The individual records ranged in yard 7 from 205 to 14 and in yard 8 from 211 to 18. The average of the two yards was 135.49 eggs. The pure-bred Leghorn dams averaged 106.88

Fig. 6. Representatives of Yard E, Oregons, 1913-14. Table XXVIII.

and their pure-bred Barred Rock dams 86.14, while their half-sister pure-bred Leghorns averaged 104.67 (see Table XIV), and Barred Rocks 120.68 (see Table II). It will be understood, of course, that the dams of the cross-breds were not the same individuals as the dams of the pure-breds, and it may have happened that the production capacity was not the same in each case. The fact, however, remains that the cross-bred pullets of the first generation were half sisters of the pure-bred Leghorn and Barred Rock pullets, being bred from the same males, and they were from the same stock on the dam's side, but from different hens. The experiment had to do, however, with inheritance in egg laying, and not with relative merits of pure-breds and crosses.

Third Generation. The first generation of cross-bred pullets were mated back to White Leghorn males, but of unknown production pedigree. The progeny is shown in yard 1 (Table XXVI), thirty-seven pullets with average of 147.16 eggs.

Fourth Generation. In the first generation of pedigreed cross-breds (Table XXVII) referred to hereafter as Oregons, twenty pullets in yard 9 averaged 220.8 eggs, ranging in production from 303 to 123. These were sired by a Leghorn male, whose dam laid 229 eggs, except in three cases where an Oregon male was used on his dam. There were three pullets from this inbred mating, namely C543, C525, C540. If these had been eliminated in the tabulation the record of the yard would have been one-half dozen more eggs a hen. Probably they should have been eliminated as in all of our experiments we sought to avoid inbreeding. The average of the dams in yard 9, was 207.45 eggs and of the sires' dams 224.42.

Fifth Generation. In the second generation of pedigreed Oregons, the records of 49 pullets are given in Table XXVIII, yard E. This yard was produced by mating daughters of cross-bred hen 250, with record of 402 eggs in two years, with a Barred Rock male, whose dam A111 produced 204 eggs in her pullet year. These 49 pullets averaged 220.78 eggs, this average exceeding the record of the dam as well as the sire's dam. The record of dams' dam was 200 the first year and 202 the second year. The high second-year record might indicate that she did not lay her full capacity in the first year. The highest record of yard E was 278 and the lowest 58. The dams of 31 of the pullets averaged 186.45 the first year. In the case of eighteen pullets, the dams are not

Fig. 7. Pen 4 of Oregons, 1914-15; record 250.2 eggs. All daughters of C521, the first 300-egg hen. See Table XXIX.

known, but they were in yard C, as well as the thirteen, with pen pedigrees. Yard C pullets were late hatched, averaging 173.66 the first year, but in the second year averaged 199.09.

Sixth Generation. Yard 4, 1914-15 (Table XXIX) gives the production of ten Oregons that are full sisters; the dam was C521, our first 300-egg hen. The average of the ten pullets was 250.2 eggs, the highest being 283 and the lowest 204, all exceeding 200 eggs. The pedigree showed dam's average 303, dam's dam 201, sires' dam 228.

Another yard of full sisters to pullets in yard 4 is not given in the tabulation on account of the pullets being late or June hatched. The record of the 13 pullets in yard 11, however, was 181.62 eggs, average, with highest 234 and lowest 99.

In Table XXX, the same year, ten pullets in yard 5 averaged 201.70. These pullets were full sisters, with dam's record of 291 eggs, dam's dam 200 eggs, sire's dam 246. In this case there was inbreeding, the dam and sire being from the same dam, but from different sires. The highest record in this yard was 241 eggs and the lowest 152.

In yard 6 (Table XXXI), the same year, were 14 pullets showing a production of 217.27 eggs, the dams averaging 211.43, dams' dams 233.5, sires' dams 286.51.

Yard J (Table XXXII) in the same year, 58 Oregons, laid an average of 219.22 eggs, with high record of 309 and low record at 77. In this

Fig. 8. Oregon hens C543 and number of eggs she laid in first year, 291.

case the highest hen should probably be eliminated because she was bred from a low line. Through an oversight the hen was included in the tabulation. Her record and breeding are discussed in another place. If her record were omitted, however, it would make a difference in the average in the yard of less than 1½ eggs a hen.

The average of the dams' of yard J was 215.24, dams' dams 217.4, sires' dams 214.31. Yard E appearing in the dam's column averaged in the preceding year 220.78, so that the dams of all yard-J pullets averaged over 200 eggs.

Seventh Generation. Only one yard in 1915-16 (shown in Table XXXIII) is available for comparison. The average of the 13 pullets in this yard was 237.77 eggs. The dams are not known in each case, but the sire's dam was C521, record 303; the average of yard M, which was made up of the best hens from yard E (see Table XXVIII) 1913-14 pullets, was 230.42 eggs. The dams' dams' average was also high, and sires' dams' average was 297.46.

To try out the pullets of this breeding under different conditions, a pen of ten was sent to the International Egg-laying Contest at the Storrs Agricultural College, Connecticut; the pen made a record of 212.2 eggs a hen. Another pen of five pullets of the same breeding was sent to the Missouri laying contest and they made a record of 232 eggs a hen. Yard 10 and the two contest pens were all of the same hatch, and of the same pedigree. After yard 10 and the two contest pens were selected, there remained of the flock of pullets 25 which were thought to be the poorest of the flock, judging from external points. It was decided to put these under test under decidedly different conditions; namely, on the writer's backyard lot, where they were kept for the full year

Fig. 9. A pen of ten Oregons, winners in International Egg-laying Contest at the Connecticut Agricultural College, 1917-18, record 235.2 eggs a hen. All from pedigreed high producers. This is the highest pen record made there in seven annual contests, representing 700 different pens.

in a house 10 by 14 feet in size, and never out of doors. These were cared for by a school boy of 13 years of age, but they got good care. They averaged 216 eggs in the year. The records of these four different pens would indicate that the breeding was responsible for their high production; also that a hen of good laying capacity will respond well to a wide variation in conditions, whether it be feeding, housing, or climate.

Eighth Generation. In 1917-18, pen 7 (Table XXXIV) containing twelve Oregon pullets averaged 246.91 eggs, with dams' average of 250.92, dams' dams 261.14, sires' dams 287.75. In the same year a pen of ten pullets of practically the same breeding won the egg-laying contest at the Storrs College, with a record of 235.2 eggs, the highest record made at that contest in seven years with ten hens, 100 pens each year.

Sixty-three pullets in yard H, 1917-18 (Table XXXV) averaged 228.57, dams' average 237.79. Dams' dams 269.44, sires' dams 269.62. Highest record 304, lowest 123.

Effect of Close Confinement. In the year 1914-15 another pen of pullets was used which were kept under continuous confinement throughout the year, and for that reason are not included in the tabulation on inheritance. The environmental conditions were entirely different. The purpose was to observe the behavior of the strain under close confinement.

There were 25 pullets in this pen, and they were kept in the same style of house as all the other pens in our breeding experiments; namely, an open-front colony house 8 by 12 feet in dimensions. They were fed the same and were of the same breeding as yard J. They were never outside of the house during the year. One of the pullets died during the year after laying 153 eggs. The remaining twenty-four averaged 228.58. The range was from 299 eggs to 147. All but four of the pullets laid over 200 eggs. They exceeded the production of yard J by about ten eggs a hen. In this pen as in yard J the highest hen was sired by low-line male D461-2. The mortality in J was 6, or 1 in 11.5, while in "shut-in" house, it was 1 in 25.

Under radically different environment this pen shows a considerable increase in production over pen J, indicating (1) that the hen is not particularly sensitive to conditions of housing, providing other factors are favorable, including abundant fresh air and exercise; (2) that the conditions under which our experimental flocks were kept, were not unduly favorable, nor responsible for the increased flock production; (3) that a hen of good laying capacity will lay well under various conditions of housing and yarding.

TABLE XXIV. ANNUAL AND BEST TWO-MONTHS PRODUCTION
Yard 7, 1909-10, Oregons.

Hen No.	Production 1st year	Best 2 months	Hen No.	Production 1st year	Best 2 months	Hen No.	Production 1st year	Best 2 months
231	129	34	243	159	35	255	93	29
232	120	32	244	142	32	256	117	37
233	104	26	245	148	32	257	126	34
234	104	32	247	168	35	258	14	5
235	143	36	248	129	32	259	136	32
236	116	31	249	144	38	260	205	45
238	85	29	250	200	40	261	118	27
239	143	34	251	168	39	262	113	27
240	102	27	252	152	33	263	195	41
241	127	34	253	91	30	264	119	33
242	139	29	254	108	24	266	131	35
Average of yard							129.94	32.09
Unidentified eggs added							131.06	

TABLE XXV. ANNUAL AND BEST TWO-MONTHS PRODUCTION
Yard 8, 1909-10, Oregons.

Hen No.	1st year	Best 2 months	Hen No.	1st year	Best 2 months	Hen No.	1st year	Best 2 months
267	139	31	278	98	26	292	133	30
268	174	45	281	130	29	293	169	45
269	101	33	282	117	25	194	130	32
270	18	14	284	113	28	295	183	40
271	101	37	286	197	40	296	95	29
272	133	39	287	156	40	297	122	33
273	176	40	288	167	38	298	179	34
274	152	31	289	132	34	299	211	43
275	123	26	290	186	49	300	161	30
276	126	33	291	98	26	302	119	34

Average of yard .... 137.96 | 33.80

Unidentified eggs added 140.36

TABLE XXVI. ANNUAL AND BEST TWO-MONTHS PRODUCTION
Yard 1, 1910-11, Oregons.

Hen No.	1st year	Best 2 months	Hen No.	1st year	Best 2 months	Hen No.	1st year	Best 2 months
A 25	134	33	A 43	176	38	A 64	157	35
A 26	68	22	A 44	28	14	A 65	154	38
A 28	168	39	A 46	102	31	A 66	201	47
A 29	161	35	A 47	205	44	A 68	139	36
A 30	168	42	A 48	138	46	A 70	195	41
A 32	139	30	A 49	71	21	A 71	161	39
A 33	110	30	A 53	81	28	A 72	108	25
A 34	152	35	A 58	165	38	A 73	137	42
A 35	163	43	A 59	86	23	A 74	166	36
A 36	187	48	A 60	177	45	A 185	176	39
A 37	160	35	A 61	257	49	321	163	46
A 39	151	42	A 62	170	42			
A 40	112	35	A 63	159	36			

Average of yard .... 147.16 | 36.43

TABLE XXVII. ANNUAL AND BEST TWO-MONTHS PRODUCTION
Yard 9, 1912-13, Oregons.

Hen No.	1st year	Best 2 months	Dam	Eggs 1st yr.	D.D.	Eggs 1st yr.	S.D.	Eggs 1st yr.
C 503	207	50	A 47	205	......	......	O 34	229
C 508	268	56	A 66	201	......	......	O 34	229
C 511	251	58	A 60	177	......	......	O 34	229
C 512	217	49	A 60	177	......	......	O 34	229
C 518	226	50	A 47	205	......	......	O 34	229
C 519	272	52	A 47	205	......	......	O 34	229
C 521	303	57	A 66	201	......	......	O 34	229
C 523	204	46	A 66	201	......	......	O 34	229
C 525	146	36	250	200	......	......	250	200
C 528	214	51	A 66	201	......	......	O 34	229
C 530	184	48	A 60	177	......	......	O 34	229
C 534	204	48	A 61	257	......	......	O 34	229
C 535	200	41	A 61	257	......	......	O 34	229
C 539	191	48	A 61	257	......	......	O 34	229
C 540	123	26	250	200	......	......	250	200
C 541	193	41	A 61	257	......	......	O 34	229
C 542	218	47	A 61	257	......	......	O 34	229
C 543	291	60	250	200	......	......	250	200
C 545	227	50	313	113	......	......	......	......
C 547	250	54	A 66	201	......	......	O 34	229
①	219.45	48.40	......	207.45	......	......		224.42
②	220.80							

① Average of yard. ② Unidentified eggs added.

TABLE XXVIII. ANNUAL AND BEST TWO-MONTHS PRODUCTION
Yard E, 1913-14, Oregons.

Hen No.	Production 1st year	Best 2 months	Dam	Eggs 1st yr.	D.D.	Eggs 1st yr.	S.D.	Eggs 1st yr.
D 624	242	50	B170	226	250	200	A 111	204
D 625	274	52	Yd. C	*	250	200	A 111	204
D 626	205	43	B 222	188	250	200	A 111	204
D 627	246	52	Yd. C	*	250	200	A 111	204
D 628	220	52	Yd. C	*	250	200	A 111	204
D 629	190	52	B 222	188	250	200	A 111	204
D 630	231	52	B170	226	250	200	A 111	204
D 631	212	49	B170	226	250	200	A 111	204
D 632	228	55	Yd. C	*	250	200	A 111	204
D 633	202	48	Yd. C	*	250	200	A 111	204
D 634	227	55	Yd C	*	250	200	A 111	204
D 635	250	53	B 213	198	250	200	A 111	204
D 636	178	54	B 222	188	250	200	A 111	204
D 637	245	56	H 27M	133	250	200	A 111	204
D 639	185	40	H 53N	168	250	200	A 111	204
D 640	206	49	B 213	198	250	200	A 111	204
D 641	212	40	Yd. C	*	250	200	A 111	204
D 643	92	55	H 16N	184	250	200	A 111	204
D 644	232	49	B 223	146	250	200	A 111	204
D 645	201	49	Yd. C	*	250	200	A 111	204
D 647	278	51	Yd. C	*	250	200	A 111	204
D 648	195	56	B170	226	250	200	A 111	204
D 649	232	49	B 177	193	250	200	A 111	204
D 650	157	42	H 27M	133	250	200	A 111	204
D 652	221	59	Yd. C	*	250	200	A 111	204
D 653	220	51	Yd. C	*	250	200	A 111	204
D 656	176	45	Yd. C	*	250	200	A 111	204
D 657	272	57	B170	226	250	200	A 111	204
D 658	249	55	Yd. C	*	250	200	A 111	204
D 660	216	51	B 223	146	250	200	A 111	204
D 662	171	58	Yd. C	*	250	200	A 111	204
D 663	270	57	B 213	198	250	200	A 111	204
D 664	254	51	B 223	146	250	200	A 111	204
D 665	220	46	B 213	198	250	200	A 111	204
D 667	216	55	B 17	148	250	200	A 111	204
D 670	263	52	Yd. C	*	250	200	A 111	204
D 671	210	47	B 213	198	250	200	A 111	204
D 672	242	54	H 16N	184	250	200	A 111	204
D 673	58	38	Yd. C	*	250	200	A 111	204
D 674	216	54	B 4	217	250	200	A 111	204
D 675	248	54	H 16N	184	250	200	A 111	204
D 676	257	55	B 4	217	250	200	A 111	204
D 678	190	53	B 4	217	250	200	A 111	204
D 679	193	49	H 27M	133	250	200	A 111	204
D 680	185	50	Yd. C	*	250	200	A 111	204
D 681	263	57	H 16N	184	250	200	A 111	204
D 682	178	44	Yd. C	*	250	200	A 111	204
D 683	269	55	B 4	217	250	200	A 111	204
232	210	51	B 223	146	250	200	A 111	204
①	216.44	51.04	....	186.45	......	200.00	......	204.00
②	220.78							

① Average of yard. ② Unidentified eggs added.

*Pen composed of daughters of hen 250. The pen average was 173.66 in first year, and 199.09 in second year. During the first year the lowest record was 116 and the highest 226. During the second year the lowest was 164 and the highest 224.

TABLE XXIX. ANNUAL AND BEST TWO-MONTHS PRODUCTION
Yard 4, 1914-15, Oregons.

| Hen No. | Production | | Dam | Eggs 1st yr. | D.D. | Eggs 1st yr. | S.D. | Eggs 1st yr. |
	1st year	Best 2 months						
E 215	283	56	C 521	303	A 66	201	B 42	228
E 216	250	59	C 521	303	A 66	201	B 42	228
E 217	254	55	C 521	302	A 66	201	B 42	228
F 218	258	50	C 521	303	A 66	201	B 42	228
E 219	280	54	C 521	303	A 66	201	B 42	228
E 220	247	55	C 521	303	A 66	201	B 42	228
E 221	204	44	C 521	303	A 66	201	B 42	228
E 223	232	51	C 521	303	A 66	201	B 42	228
E 225	224	50	C 521	303	A 66	201	B 42	228
E 227	228	50	C 521	303	A 66	201	B 42	228
①	246.00	52.40	......	303.00	....	201.00	....	228.00
②	250.20							

① Average of yard. ② Unidentified eggs added.

TABLE XXX. ANNUAL AND BEST TWO-MONTHS PRODUCTION
Yard 5, 1914-15, Oregons.

| Hen No. | Production | | Dam | Eggs 1st yr. | D.D. | Eggs 1st yr. | S.D. | Eggs 1st yr. |
	1st year	Best 2 months						
E 200	200	56	C 543	291	250	200	B 8	246
E 201	192	52	C 543	291	250	200	B 8	246
E 202	162	52	C 543	291	250	200	B 8	246
E 204	160	47	C 543	291	250	200	B 8	246
E 205	239	50	C 543	291	250	200	B 8	246
E 206	152	45	C 543	291	250	200	B 8	246
E 207	193	53	C 543	291	250	200	B 8	246
E 208	126	48	C 543	291	250	200	B 8	246
E 209	241	48	C 543	291	250	200	B 8	246
E 211	205	53	C 543	291	250	200	B 8	246
①	197.00	50.4	....	291.00	......	200.00	......	246.00
②	201.70							

① Average of yard. ② Unidentified eggs added.

TABLE XXXI. ANNUAL AND BEST TWO-MONTHS PRODUCTION
Yard 6, 1914-15, Oregons.

| Hen No. | Production | | Dam | Eggs 1st yr. | D.D. | Eggs 1st yr. | S.D. | Eggs 1st yr. |
	1st year	Best 2 months						
E 232	145	50	A 66	201	....	......	C 543	291
E 233	212	48	A 66	201	....	......	C 543	291
E 234	286	57	C 507	215	A 27	240	C 543	291
E 235	253	56	C 507	215	A 27	240	C 543	291
E 236	156	52	A 66	201	....	......	C 543	291
E 237	235	48	A 66	201	....	......	C 543	291
E 238	239	54	B 42	228	....	......	C 543	291
E 239	159	47	A 66	201	....	......	C 543	291
E 240	257	53	C 507	215	A 27	240	C 543	291
E 241	212	52	C 507	215	A 27	240	C 543	291
E 242	161	47	C 507	215	A 27	240	C 543	291
E 243	219	53	C 547	250	A 66	201	B 42	228
E 244	177	44	A 66	201	....	......	C 543	291
E 246	245	51	A 66	201	....	......	C 543	291
①	211.14	50.86	....	211.43	....	233.50	......	286.51
②	217.27							

① Average of yard. ② Unidentified eggs added.

TABLE XXXII. ANNUAL AND BEST TWO-MONTHS PRODUCTION
Yard J, 1914-15, Oregons.

Hen No.	Production		Dam	Eggs 1st yr.	D.D.	Eggs 1st yr.	S.D.	Eggs 1st yr.
	1st year	Best 2 months						
E 101	178	51	Yd. E	......	....	......	A 66	201
E 102	241	56	Yd. E	......	....	......	A 66	201
E 103	162	37	C 539	191	A 61	257	B 8	246
E 104	263	51	Yd. E	......	....	......	A 66	201
E 105	181	46	Yd. E	......	....	......	A 66	201
E 106	238	52	Yd. E	......	....	......	A 66	201
E 107	213	53	C 490	231	A 61	257	B 8	246
E 108	212	46	Yd. E	......	....	......	A 66	201
E 109	200	46	Yd. E	......	....	......	A 66	201
E 110	233	49	Yd. E	......	....	......	A 66	201
E 112	244	47	Yd. FX				A 27	240
E 113	224	51	C 539	191	A 61	257	B 8	246
E 114	262	54	C 470	211	A 47	205	H12N	144
E 115	301	58	B 42	228	....	......	C 543	291
E 116	149	34	Yd. FX			.... .	A 27	240
E 117	176	40	Yd. E	......	....	......	A 66	201
E 118	258	48	B 42	228	....	......	C 543	291
E 119	245	54	C 508	268	A 66	201	C 543	291
E 120	77	30	Yd. E	......	....	......	A 66	201
E 121	180	48	C 528	214	A 66	201	B 42	228
E 122	199	45	Yd. E	......	....	......	A 66	201
E 123	293	56	Yd. E	......	....	......	A 66	201
E 124	238	49	C 537	219	A 51	209	C 543	291
E 125	188	46	B 8	246	250	200	B 42	228
E 126	196	54	C 514	214	A 45	215	C 543	291
E 127	274	56	C 503	207	A 47	205	H12N	144
E 128	150	45	C 425	235	....	......	B 42	228
E 129	233	49	Yd. E	......	....	......	A 66	201
E 130	282	57	Yd. E	......	....	......	A 66	201
E 131	162	41	C 539	191	A 61	257	B 8	246
E 132	261	47	Yd. E	......	....	......	A 66	201
E 133	228	50	Yd. E	......	....	......	A 66	201
E 134	182	46	Yd. E	......	....	......	A 66	201
E 135	230	46	B 222	188	250	200	B 42	228
E 136	285	59	Yd. E	......	....	......	A 66	201
E 138	309	57	C 584	119	O 34	229	H12N	144
E 139	191	40	C 534	204	A 61	257	C 543	291
E 141	228	47	Yd. E	......	....	......	A 66	201
E 143	155	45	C 547	250	A 66	201	B 42	228
E 144	154	46	Yd. E	......	....	......	A 66	201
E 145	195	44	Yd. E	......	....	......	A 66	201
E 146	208	47	Yd. E	......	....	......	A 66	201
E 147	241	58	C 503	207	A 47	205	H12N	144
E 148	181	44	Yd. E	......	....	......	A 66	201
E 149	159	47	Yd. E	......	....	......	A 66	201
E 150	191	53	Yd. E	......	....	......	A 66	201
E 152	213	53	Yd. E	......	....	......	A 66	201
E 153	213	53	Yd. E	......	....	......	A 66	201
E 154	280	57	Yd. E	......	....	......	A 66	201
E 155	253	51	Yd. E	......	....	......	A 66	201
E 156	284	54	C 518	226	A 47	205	H12N	144
E 157	266	54	C 512	217	A 60	177	A 27	240
E 159	171	51	Yd. E	......	....	.... .	A 66	201
E 163	194	46	Yd. E	......	....	......	A 66	201
E 165	236	54	C 537	219	A 51	209	C 543	291
E 166	147	43	C 547	250	A 66	201	B 42	228
1554	270	56	D 618	188	Yd. C	......	B 170	226
1604	223	46	C 481	239	Yd. J	......	H12N	144
①	217.24	49.19	....	215.24	....	217.40	......	214.31
②	219.22							

① Average of yard.   ② Unidentified eggs added.

TABLE XXXIII. ANNUAL AND BEST TWO-MONTHS PRODUCTION
Yard 10, 1915-16, Oregons.

Hen No.	Production		Dam	Eggs 1st yr.	D.D.	Eggs 1st yr.	S.D.	Eggs 1st yr.
	1st year	Best 2 months						
F 566	286	58	D 269	276	B 42	228	C 516	267
F 567	271	56	Yd. M	*	....	......	C 521	303
F 568	222	52	Yd. M	*	... .	......	C 521	303
F 569	208	50	Yd. M	*	....	......	C 521	303
F 571	251	51	Yd. M	*	....	......	C 521	303
F 572	245	48	Yd. M	*	....	......	C 521	303
F 574	233	44	Yd. M	*	....	......	C 521	303
F 575	156	40	Yd. M	*	....	... ..	C 521	303
F 576	226	53	Yd. M	*	....	... ..	C 521	303
F 577	226	54	Yd. M	*	....	......	C 521	303
F 578	200	46	D 614	272	B 4	217	C 516	267
F 579	256	49	Yd. M	*	....	......	C 521	303
F 580	267	53	Yd. M	*	....	......	C 521	303
①	234.39	50.31	....	274.00	....	222.50	....	297.46
②	237.77							

① Average of yard.   ② Unidentified eggs added.
* Pen of high producing birds laying from 190 to 278.

TABLE XXXIV. ANNUAL AND BEST TWO-MONTHS PRODUCTION
Yard 7, 1917-18, Oregons.

Hen No.	Production		Dam	Eggs 1st yr.	D.D.	Eggs 1st yr.	S.D.	Eggs 1st yr.
	1st year	Best 2 months						
H 589	270	56	F 567	271	Yd. M	......	C 521	303
H 590	284	55	E 827	203	C 488	268	C 521	303
H 591	244	51	F 572	245	Yd. M	......	C 521	303
H 592	265	50	E 220	247	C 521	303	D 269	276
H 593	189	43	F 566	286	D 269	276	C 521	303
H 594	240	49	F 568	222	Yd. M	......	C 521	303
H 595	218	48	F 808	262	E 217	254	E 123	293
H 596	251	45	F 813	206	E 216	250	E 123	293
H 597	210	51	F 566	286	D 269	276	C 521	303
H 598	251	54	F 816	265	Yd. 7	......	E 123	293
H 599	274	50	F 281	268	Yd. M	......	A 60	177
H 600	224	53	C 547	250	A 66	201	C 521	303
①	243.33	50.41	....	250.92	....	261.14	....	287.75
②	246.91							

① Average of yard.   ② Unidentified eggs added.

TABLE XXXV. ANNUAL AND BEST TWO-MONTHS PRODUCTION
Yard H, 1917-18, Oregons

Hen No.	Production 1st year	Production Best 2 months	Dam	Eggs 1st yr.	D.D.	Eggs 1st yr.	S.D.	Eggs 1st yr.
H 701	177	50	C 547	250	A 66	201	C 521	303
H 703	283	54	E 220	247	C 521	303	D 269	276
H 704	191	55	C 547	250	A 66	201	C 521	303
H 705	191	47	C 547	250	A 66	201	C 521	303
H 706	208	51	Yd. 13	....	C 488	268	A 60	177
H 707	243	51	E 827	203	Yd. M	......	C 521	303
H 708	200	45	F 567	271	Yd. M	......	C 521	303
H 709	229	55	Yd. 8	....	......	......	B 14	215
H 710	170	50	E 229	241	C 521	303	C 521	303
H 713	276	58	Yd. 13	....	......	......	A 60	177
H 714	221	51	E 285	234	C 521	303	D 269	276
H 715	251	54	E 285	234	C 521	303	D 269	276
H 716	203	48	Yd. 13	....	......	......	A 60	177
H 717	252	52	E 282	233	C 521	303	D 269	276
H 718	223	56	Yd. 9	....	C 521	303	E 178	299
H 719	234	54	Yd. 13	....	......	......	A 60	177
H 721	226	50	F 575	156	Yd. M	......	C 521	303
H 722	256	51	Yd. 13	....	......	......	A 60	177
H 723	236	54	E 827	203	C 488	268	C 521	303
H 724	177	44	F 554	234	C 521	303	E 178	299
H 725	236	54	Yd. 8	....	......	......	B 14	215
H 726	231	54	F 575	156	Yd. M	......	C 521	303
H 728	269	55	E 827	203	C 488	268	C 521	303
H 729	173	52	Yd. 9	....	C 521	303	E 178	299
H 731	230	51	Yd. 9	....	C 521	303	E 178	299
H 732	207	50	F 575	156	Yd. M	......	C 521	303
H 733	192	52	F 555	132	C 521	303	E 178	299
H 734	175	56	G 515	274	C 521	303	C 590	193
H 735	225	53	E 229	241	C 521	303	C 521	303
H 736	235	55	F 555	132	C 521	303	E 178	299
H 737	222	58	Yd. 9	....	C 521	303	E 178	299
H 740	227	57	Yd. 8	....	......	......	B 14	215
H 741	204	50	F 554	234	C 521	303	E 178	299
H 742	260	49	Yd. 13	....	......	......	A 60	177
H 743	170	55	Yd. 9	....	C 521	303	E 178	299
H 745	213	48	E 157	266	C 512	217	E 123	293
H 746	216	50	F 916	213	B 14	215	E 123	293
H 747	244	52	D 328	215	B 13	206	E 123	293
H 749	229	51	F 916	213	B 14	215	C 521	303
H 750	261	56	F 566	286	D 269	276	C 521	303
H 751	253	54	E 123	293	Yd. E	......	E 123	293
H 752	202	47	E 136	285	Yd. E	......	E 123	293
H 753	209	53	Yd. 9	....	C 521	303	E 178	299
H 755	228	51	E 130	282	Yd. E	......	E 123	293
H 756	192	47	E 123	293	Yd. E	......	E 123	293
H 757	213	51	F 568	222	Yd. M	......	C 521	303
H 758	203	49	F 176	242	B 14	215	E 123	293
H 759	271	61	F 846	265	Yd. 7	......	E 123	293
H 760	291	59	F 566	286	D 269	276	C 521	303
H 761	196	56	E 130	282	Yd. E	......	E 123	293
H 762	304	58	E 112	244	Yd FX	......	E 123	293
H 763	229	49	F 873	250	B 8	246	S 123	293
H 764	189	47	D 328	215	B 13	206	E 123	293
H 767	226	55	D 269	276	B 42	228	C 521	303
H 768	247	49	D 269	276	B 42	228	C 521	303
H 769	255	54	Yd. 9	....	C 521	303	E 178	299
H 770	258	58	Yd. 9	....	C 521	303	E 178	299
H 771	269	55	F 281	268	Yd. M	......	A 60	177
H 772	189	54	G 515	274	C 521	303	C 590	193
H 773	270	51	Yd. 13	....	C 521	303	A 60	177
H 774	123	38	Yd. 13	....	......	......	A 60	177
H 775	228	52	Yd. 13	....	......	......	A 60	177
H 776	225	59	E 515	245	C 56	210	A 27	240
①	224.38	52.30	....	237.79	....	269.44	....	269.62
②	228.57							

① Average of yard.   ② Unidentified eggs added.

SUMMARIES

Barred Plymouth Rocks. A summary is given in Table XXXVI of the Barred Plymouth Rocks, showing egg records for first twelve months of laying of the pullets, their dams' dams, sires' dams where known for the different years. This gives the record of 517 pullets in their first year, and the average for all of the years, for eight generations, including the foundation stock, together with the records of ancestry. Beginning with the foundation stock in 1908-09, and ending with the eighth generation, there is an increase in the production per hen of 128.49 eggs, or a percentage increase of 149.16. If we make the starting point 1909-10 in view of the fact that the pullets of the previous year were selected from different breeders and were not reared at the Station as all subsequent pullets were, we find the increase to be 93.95 eggs, or 77.78 percent.

Yard 15 (Table III), though not pedigreed, were from hens of the first year after being culled. That is to say, one-third of the poorest of the first year were culled out, yard-15 pullets being therefore from the best two-thirds of the flock. This may have had some influence in raising the production of yard 15. The males were of unknown pedigree, but it may have "happened" that they were from good layers. The pullets in this yard, in addition, were of a uniform age, and came to laying maturity at a favorable time, and in this respect had the advantage to some extent over yards of later years.

The first records of pedigreed pullets are given in yard 6, 1912-13 (Table IV). The pullets of this yard were from dams with an average record of 210.94, and sires' dams 194.79.

It is seen that there is a considerable increase in egg yield in the first-generation pullets from high-producing pedigreed stock. The in-

TABLE XXXVI. SUMMARY OF FIRST YEAR'S PRODUCTION OF
BARRED PLYMOUTH ROCKS (IN AVERAGE PER HEN)

Yard	Year	No. of birds	Pullets	Unidentified eggs added	Best two months	Dams	Dams' dams	Sires' dams
4 and 5..	1908-09	92	82.67	86.14	31.29	....	....	....
6............	1909-10	28	117.64	120.68	33.71	....	....	....
15............	1910-11	42	161.78	164.28	38.48		....	
6............	1912-13	38	174.13	178.16	44.74	210.94	....	194.79
7............	1912-13	37	177.43	180.97	45.22	181.37	202.44	185.96
8............	1912-13	33	171.69	174.18	44.30	196.76	....	183.85
R............	1913-14	52	178.75	185.00	46.88	193.86	215.05	220.52
C............	1913-14	51	169.41	175.59	46.57	182.32	205.75	224.57
L............	1914-15	33	176.36	185.78	45.27	181.33	208.71	216.19
P............	1914-15	38	184.06	189.01	46.69	213.84	214.81	231.68
18............	1916-17	11	194.45	201.90	51.18	247.36	197.44	223.13
E............	1917-18	62	202.40	214.63	47.19	200.23	229.47	244.58
Average..............517			158.57	164 09	42.25	198.11	212.45	217.29

crease is not so great in later generations. Where the increase from unselected stock to pedigreed stock of the first generation is about 56 eggs the increase from the first generation of pedigreed stock to the fourth generation of pedigreed pullets is about 24 eggs. While the increases in subsequent generations are satisfactory, the largest increase is shown in the first generation of selected stock. The influence of the immediate parent is apparently more pronounced than that of other

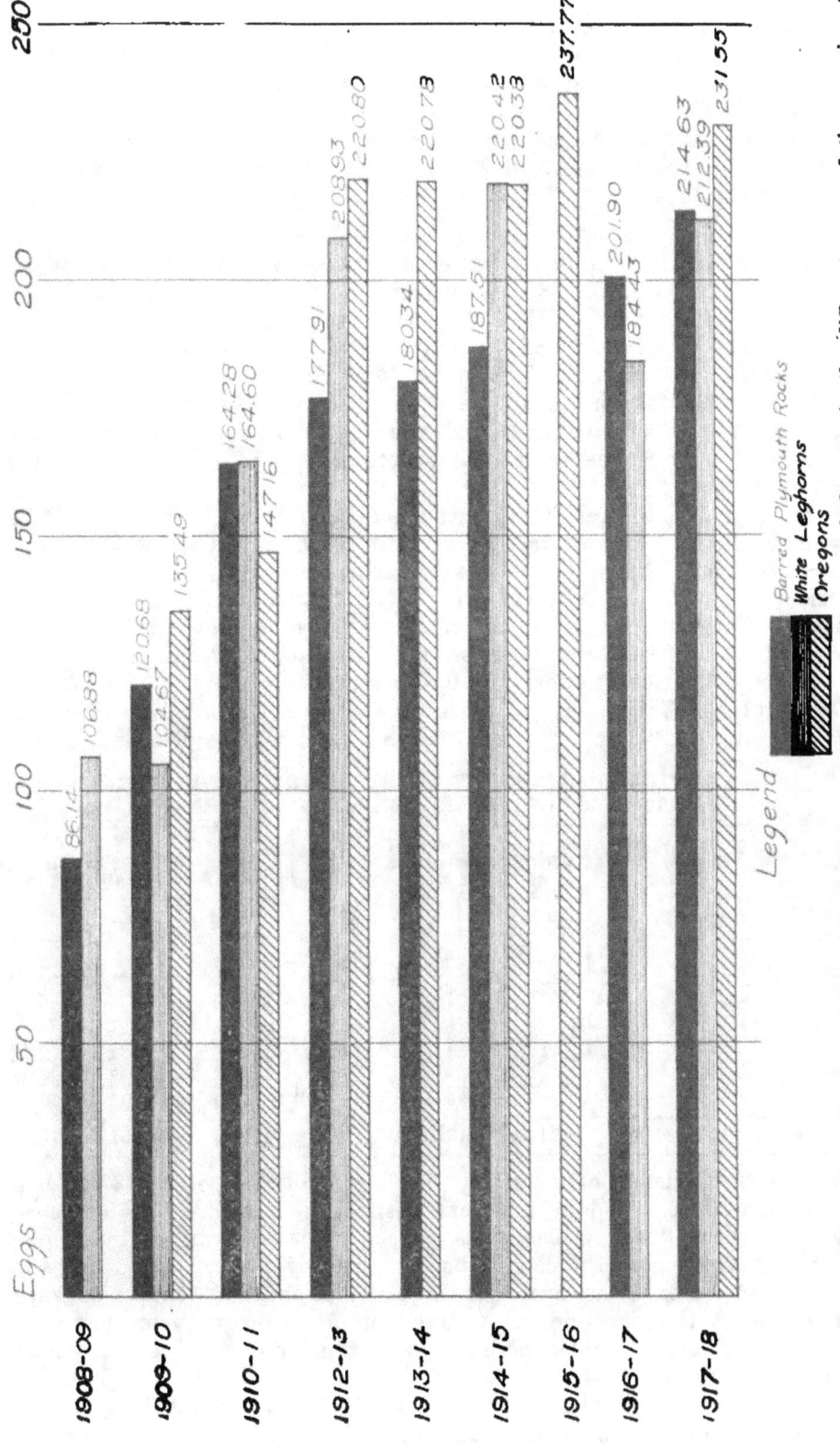

Fig. 10. Average first year's production of Barred Plymouth Rocks, Leghorns, and Oregons in the different years of the experiment, showing increase in production in different generations as a result of selection of breeding stock on basis of first year's production records.

ancestors. On the other hand, we see from the records that the more high-record ancestors there are, the higher is the flock average. It is expected, of course, that as the maximum possible hen production is approached, whatever it is, the rate of increase must necessarily be lower. In other words, as production rises beyond the mean of the breed or race, there is a stronger pull backward and increases will naturally be slower.

White Leghorns. A summary for the White Leghorns is given in Table XXXVII. The total number of hens was 386. The average of the foundation flock was 106.88 eggs a hen. In the last year yards F and G (Tables XXII and XXIII), 116 hens, averaged 212.39 a hen, which is a percentage increase of 98.72. In the third year, with only ten hens, the average was high for unselected hens, but the fact that in this yard hen A27 of high record was "found" accounts for some of the increase. These hens were not pedigreed but were from the foundation stock in the second year after the poorest layers of the first year had been culled out.

In the main, the results with the Leghorns agree with those of the Plymouth Rocks. In the first generation of pedigreed stock there was a large increase. In the second year of pedigreed stock, there was also a considerable increase. After that the increase was not particularly noteworthy. The variation in the increase in different years is considerable, and does not conform to any statistical conception of breeding.

Probably various factors account for these differences. In yard 3, 1914-15 (Table XVII), the influence of the male was plainly apparent. The breeding quality or prepotency of this male will be discussed in the

TABLE XXXVII. SUMMARY OF FIRST YEAR'S PRODUCTION OF WHITE LEGHORNS (IN AVERAGE PER HEN)

Yard	Year	No. of birds	Pullets	Unidentified eggs added	Best two months	Dams	Dams' dams	Sires' dams
4 and 5	1908-09	50	106.14	106.88	32.54	....	....	....
9	1909-10	21	104.67	104.67	30.71	....	....	....
1	1910-11	10	164.60	164.60	41.30	....	....	....
9	1912-13	14	207.85	208.93	47.64	223.85	....	229.00
3	1914-15	16	227.00	230.12	51.19	242.19	222.33	233.13
O	1914-15	42	212.86	216.72	49.24	236.97	221.68	237.25
P	1916-17	49	187.55	188.96	47.61	209.74	247.27	251.71
Q	1916-17	57	169.84	172.65	44.77	211.75	241.56	253.74
15	1916-17	11	215.00	225.27	51.63	227.64	245.09	242.18
F	1917-18	60	218.21	222.22	47.78	235.42	239.78	302.00
G	1917-18	56	199.63	201.85	48.57	234.50	247.95	236.81
Average		386	181.71	184.25	44.76	224.69	238.52	256.33

report on the influence of the male. There was in this pen a 302-egg hen and several sisters and half sisters with high records. The decrease shown in yards P and Q has already been referred to as probably due largely to environmental conditions. It will be noticed also that the production of the dams averaged lower than that of other pedigreed yards, though the sires' dams averaged high. In every year, however, there was a considerable increase over the production of the non-pedigreed flock.

A flock of sixty pullets in yard F, 1917-18 (Table XXII) gave the very satisfactory average of 222.22 eggs a hen. These pullets were all sired by males from a dam with record of 302 eggs and grandson of the sire of yard-3 pullets.

The production of the pullet progeny, it will be seen, averaged lower than that of either the dam, dams' dams, or sires' dams in all cases. But the difference is less between the pullets and dams than between the pullets and sires' dams.

The best two months' production shows an increase of 48 percent from the first year to the last year. The rate of increase in each year of pedigreed progeny, as in the case of the Plymouth Rocks, does not bear a close relation to the increase in number of generations of pedigreed ancestors.

Cross-breds or Oregons. A summary of the cross-breds or Oregons in Table XXXVIII shows that the second generation, yards 7 and 8 (Tables XXIV and XXV), averaged 135.49 eggs, and the seventh or last generation, yards 7 and H, averaged 231.55 eggs a hen, an increase of 96.06 eggs or a percentage increase of 70.89 (Tables XXXIV and XXXV). If we figure from the foundation stock of White Leghorns and Barred Plymouth Rocks, the increase is over 150 percent.

The increase in the best two months, from the first generation of crosses to the last generation of Oregons is 58.05 percent, giving additional evidence that high-laying capacity has been inherited.

There was a very marked increase from the unselected stock to the first generation of pedigreed stock, and the increase in subsequent generations is considerable, though not so marked. This agrees with the results secured with Barred Plymouth Rocks and White Leghorns.

On the whole the flock increases are much greater for the Oregons than for the Leghorns or Plymouth Rocks, if we calculate from the foundation stock. The first cross shows a considerable increase over both parent breeds. In the absence of selection, the only explanation for this increase is that there was an increase in the vigor, which was responsible for the increased production. The largest increase, however,

TABLE XXXVIII. SUMMARY OF FIRST YEAR'S PRODUCTION OF OREGONS (IN AVERAGE PER HEN)

Yard	Year	No. of birds	Pullets	Unidentified eggs added	Best two months	Dams	Dams' dams	Sires' dams
7	1909-10	33	129.94	131.06	32.09			
8	1909-10	30	137.96	140.36	33.80			
1	1910-11	37	147.16	147.16	36.43			
9	1912-17	20	215.45	220.80	48.40	207.45		224.42
E	1913-14	49	216.44	220.78	51.04	186.45	200.00	204.00
4	1914-15	10	246.00	250.20	52.40	303.00	201.00	228.90
5	1914-15	10	197.00	201.70	50.40	291.00	200.00	246.00
6	1914-15	14	211.14	217.27	50.86	211.43	233.50	286.51
J	1914-15	58	217.24	219.22	49.19	215.24	217.40	214.31
10	1915-16	13	234.39	237.77	50.31	274.00	222.50	297.43
7	1917-18	12	243.33	246.91	50.41	250.92	261.14	287.75
H	1917-18	63	224.38	228.57	52.30	227.79	269.44	269.62
Average		349	197.58	201.05	45.95	227.51	226.15	240.92

came in subsequent generations of pedigreed stock, where the record shows high production of the ancestors. In this latter respect also, the results agree with those of the pure Leghorns and Barred Plymouth Rocks, the immediate parents apparently exercising a greater influence on the production of the flock than other ancestors.

In the first generation of pedigreed Oregons, the flock average was more than the average of the dams and slightly less than the average of the sires' dams. In the next year the flock averaged 220.78 eggs, while the dam's average was 186.45, dams' dam 200, and sire's dam 204. In this case, there was an actual increased production over the parent stock on both sides. This happened in one other case; that of yard J, 1914-15 (Table XXXII), the pullets, 58 in number, averaging 219.22 eggs, dams 215.24, dams' dams 217.4, and sire's dams 214.31. Here we have an actual increase in production over the apparent capacity of the ancestors.

Taking the average of all flocks and hens where the breeding stock is selected among the highest producers of the flock, their pullet progeny has a lower average flock record than the average of the ancestors.

There are many cases, however, in our records, where the individual production of the hen is greater than the ancestors, even though the latter are high producers. In yard 9, 1912-13 cited above (Table XXVII), there are six hens in twenty with records exceeding the records of both dam and sire's dam, the highest record being 303, and dam's record 201 and sire's dam 229. The second highest was 291 eggs with dam's record of 200 and sire's record the same.

In yard E 1913-14 (Table XXVIII) is the rather remarkable record of 49 pullets, with average production considerably in excess of the production of either the dam, dam's dams, or sire's dam. The same male was sire of all the pullets. On the other hand, in yard 5 (Table XXX), ten pullets averaged 201.7, while the dam laid 291, dam's dam 200, sire's dam 246.

In yard J (Table XXXII), containing 58 pullets, the average production is higher than that of the dam, dams' dam, or sires' dams. This is another case in which the production was higher than that of high-record ancestry. In this yard, however, there were six pullets, from a low-record sire's dam and high dam, except in one case where the dam was a low hen with production of 119 eggs. From this low-record hen and low-record sire's dam, the highest egg record of the yard was secured; namely, 309 eggs. The pedigree of the sire's dam, whose record was 144 eggs, extends farther back through several generations of very low producers, as shown in Table XLV.

From this table it might be proved, both that high production is and that it is not inherited. If we take the average of all the pullet flock and the average of the ancestors, it is clear that high production is inherited. But if certain individuals be selected from the table, it could be shown that there was both transmission and non-transmission of laying qualities. Certain individuals apparently inherited high production from high producers, and certain other individuals apparently inherited low production. But when the average of all the flock is taken, in connection with the production of former generations of the strain, there can be no question that the factor of high production was inherited.

If there was lack of vigor in the parent stock and the first cross restored the vigor, then the production of the first generation of crosses, which was 135.49 eggs, should represent their actual inherited egg-laying capacity, and the increase from the first generation to the seventh of the crosses or "Oregons" represents the effect of selection of breeding stock, based on high production records. In that case, crossing alone did not produce high-record layers, the greatest increase coming from subsequent selection.

While crossing evidently increased the vigor and to that extent increased production, it is not assumed that the production of the first crosses was greater than may be secured from pure breds of good vigor; but in subsequent generations when trap-nest selection was practiced, a much greater increase was secured. It would appear that the opinion often expressed, that crossing usually gives good results in the first generation, but that in subsequent crosses there is reversion or a decrease, is not necessarily true. The increase in production of the pure White Leghorns and Barred Plymouth Rocks was secured not by maintaining merely breed purity, but by selection within the breed, or the use of breeding stock with high individual production records. Nor did we find that crossing the crosses brought about deterioration, but by similar selection of breeding stock we secured, on the contrary, a higher production than from the first generation of crosses. Inbreeding was avoided, because of the danger of defeating the purpose we had in view in crossing, that of increasing the vigor.

SUPPLEMENTARY EXPERIMENTS

In 1914 the Station placed with the poultry department of the Oregon State Hospital 100 pullets from yard E, 1913-14, Oregons. Their date of hatch was March 15. The State institution also secured eggs from the same flock from which 80 additional pullets were reared, which were hatched June 24. These two flocks were kept under conditions of housing and yarding similar to those at the Experiment Station and the same feeding system was followed. This work at the Oregon State Hospital was made possible through the cooperation of the superintendent, Dr. R. E. Lee Steiner. The record work was supervised by the head of the Poultry department of the Experiment Station, who also selected the breeding stock. The Poultry department of the State Hospital was put in direct charge of C. M. Wilcox, a graduate assistant of the Station's Poultry department. After Mr. Wilcox's death in the war service, W. H. Hart and A. D. Zinser had charge.

TABLE XXXIX. FIRST YEAR'S PRODUCTION 1914-15, OREGONS, OREGON STATE HOSPITAL

Pullet No.	Yard 1 first year	Pullet No.	Yard 2 first year	Pullet No.	Yard 3 first year
12	275	48	278	82	281
16	258	46	267	75	276
11	253	57	263	90	266
14	252	47	254	94	257
7	250	55	250	100	253
5	250	39	248	91	252
23	245	44	246	71	248
19	245	45	245	69	245
15	244	58	244	74	244
6	235	61	244	83	240
30	227	63	238	70	239
26	225	66	238	99	239
31	225	49	237	92	238
32	223	42	237	89	237
22	223	52	235	97	234
20	222	60	234	77	225
10	220	62	228	96	220
29	220	43	226	87	215
9	212	59	219	78	214
27	207	64	214	81	213
4	202	51	213	72	211
3	201	35	211	68	209
8	201	53	210	93	207
24	200	38	206	76	206
1	194	50	201	85	205
33	193	67	198	73	195
2	186	54	196	80	195
28	183	36	193	86	188
17	181	40	190	95	176
18	180	65	183	79	170
21	177	37	185	98	162
25	163	56	177	84	156
13	161				

Average................216.15 Average................225.41 Average................222.38

Of the 100 March-hatched pullets 97 completed the full laying year, while 75 of the 80 June-hatched pullets finished the laying year. The production of the two flocks is shown in tables XXXIX and XL. The first flock, which occupied yards 1, 2, and 3, averaged in the first year 221.26, with high record of 281 and low record of 156. The 75 June-hatched pullets in yards 4 and 5 averaged in the first year 219.41, with high record of 303 and low record of 147 eggs.

The pullets of yards 1, 2, and 3 were from yard-E dams, which averaged 220.78 eggs. They were not individually pedigreed, however. It will be noted that the sires' dam A66 was the dam of C521, record 303 eggs. The dams' dams were yard C, record 173.66 eggs in the first year and 199.09 in the second, the low first-year record being due

TABLE XL. FIRST YEAR'S PRODUCTION 1914-15, OREGONS, OREGON STATE HOSPITAL

Pullet No.	Yard 4 first year	Dam	First year	Pullet No.	Yard 5 first year	Dam	First year
102	303	D 681	263	174	286	D 670	263
131	296	D 657	272	159	278	D 683	269
105	292	D 683	269	146	267	D 684 or D626	......
103	291	D 674	216				
112	284	D 625	274	142	263	D 637	245
114	272	D 637	245	143	260	D 625	274
136	257	D 670	263	180	248	D 683	269
128	253	D 627	246	144	247	D 635	250
133	253	D 670	263	145	247	D 680	185
101	248	D 624	242	172	243	D 671	210
123	245	D 637	245	147	238	D 663	270
120	242	D 637	245	167	234	D 665	220
124	242	D 627	246	161	231	D 633 or D625	......
130	238	D 656	176				
109	237	D 680	185	173	230	D 684 or D626	......
108	233	D 657	272				
118	230	D 653	220	155	227	D 650	157
125	223	D 684 or D626	......	150	222	D 678	190
				149	217	D 662	171
106	221	D 636	178	152	217	D 679	193
119	218	D 625	274	163	217	D 631	212
135	212	D 670	263	171	216	D 650	157
126	211	D 675	248	148	213	D 671	210
113	210	D 624	242	154	213	D 639	185
129	208	D 675	248	179	207	D 671	210
134	208	D 671	210	141	202	D 624	242
115	207	D 629	191	168	197	D 631	212
104	207	D 678	190	166	194	D 665	220
117	205	D 655	205	156	191	D 681	263
122	200	D 625	274	178	191	D 660	216
111	197	D 662	171	160	189	D 684 or D626	......
140	195	D 684 or D626	......	170	188	D 649	232
110	194	D 629	191	153	178	D 635	250
116	190	D 679	193	175	170	D 663	270
127	185	D 679	193	176	170	D 629	191
138	178	D 679	193	151	167	D 672	242
137	173	D 625	274	177	164	D 636	178
107	167	D 635	274	157	151	D 631	212
121	157	D 633	202	165	147	232	210
132	154	D 684	218				
Average	224.00		219.85	Average	214.44		221.19

TABLE XLI.　FIRST YEAR'S PRODUCTION 1915-16, OREGONS, OREGON STATE HOSPITAL

Pullet No.	Yard 6 first year	Dam	First year	Pullet No.	Yard 6 first year	Dam	First year
317	302	14	252	332	232	8	201
285	284	16	258	366	232	100	253
423	282	76	206	421	232	85	205
217	281	91	252	447	232	61	244
234	281	81	213	432	231	69	245
433	280	16	258	287	230	32	223
369	279	1	194	398	229	71	248
407	277	31	225	451	229	15	244
224	276	63	238	314	227	50	201
341	275	79	170	413	227	26	225
203	273	82	281	209	226	89	237
278	272	93	207	275	226	2	186
215	265	75	276	364	226	54	196
291	265	95	162	367	226	11	253
308	264	76	206	281	225	22	223
411	264	96	220	252	225	87	215
435	264	59	219	213	224	100	253
343	263	25	163	309	224	31	225
402	263	82	281	391	224	27	207
261	262	40	267	406	224	22	223
452	262	12	275	300	223	44	246
420	261	87	215	331	223	36	193
259	259	94	257	254	223	94	257
211	257	82	281	339	223	1	194
294	256	12	275	246	222	23	245
443	256	59	219	225	221	58	244
276	254	71	248	279	221	78	214
256	253	74	244	438	221	91	252
271	253	93	207	257	220	20	222
292	253	57	263	289	220	76	206
320	253	94	257	297	218	16	258
340	251	11	253	418	218	27	207
251	250	12	275	210	217	92	238
201	248	23	245	266	217	14	252
221	248	82	281	359	217	26	225
375	248	15	244	319	216	75	276
307	247	68	209	404	216	46	190
349	247	81	213	441	216	4	202
429	247	64	214	207	215	26	225
361	246	16	258	267	215	13	161
448	246	48	278	286	215	77	225
449	246	72	211	365	215	35	211
229	245	82	281	370	215	83	240
274	245	45	245	394	215	5	250
227	245	87	215	299	214	69	245
395	245	42	237	330	214	35	211
337	244	8	201	345	214	10	220
235	243	71	248	204	213	5	250
218	242	93	207	306	213	51	213
219	242	28	183	444	213	67	198
226	242	90	266	352	212	9	212
327	242	82	281	321	211	23	245
212	239	68	209	368	210	42	237
293	239	75	276	233	209	75	276
344	239	98	176	255	209	91	252
371	239	59	219	405	209	42	237
236	233	87	215	250	208	67	198
242	237	8	201	419	208	46	190
357	237	1	194	237	206	57	263
231	236	77	225	315	206	15	244
283	236	59	219	335	206	39	248
362	236	76	206	355	206	75	276
220	235	81	213	414	206	46	190
393	235	90	266	346	204	67	198
228	234	16	258	351	203	65	188
272	234	75	276	303	202	13	161
347	234	77	225	354	202	45	245
342	233	45	245	442	202	44	246
428	233	6	235	268	201	14	252
273	232	11	253	392	201	13	161
277	232	69	245	403	201	16	258
				445	201	13	161

TABLE XLI.—*Continued.*

Pullet No.	Yard 6 first year	Dam	First year	Pullet No.	Yard 6 first year	Dam	First year
310	199	53	210	205	180	39	248
422	199	44	246	376	178	72	211
334	198	86	188	245	177	99	239
416	198	67	198	270	176	53	210
238	197	38	206	358	176	90	266
263	196	8	201	248	173	52	235
401	196	91	252	208	172	69	245
232	195	8	201	440	170	40	267
280	195	22	223	425	168	85	205
328	195	15	244	312	166	35	211
329	195	81	213	426	164	99	239
434	195	85	205	377	163	43	226
239	194	77	225	288	161	58	244
264	194	31	225	504	159	51	213
350	193	38	206	313	159	23	245
318	192	20	222	717	159	76	206
269	191	76	206	260	157	53	210
222	189	49	237	372	152	59	219
296	189	94	257	243	149	90	266
333	188	42	237	262	141	59	219
427	187	82	281	298	137	51	213
453	187	50	201	374	135	8	201
240	184	38	206	202	131	8	201
265	184	14	252	431	131	33	193
305	184	60	234	446	126	48	278
409	183	11	253	Average	218.78		229.17
408	182	58	244				

TABLE XLII. SUMMARY OF FIRST AND SECOND YEARS OF 77 DAMS OF YARD SIX AND THEIR PULLETS

Groups	Dams			Dam's Dams	Daughters			
	No. of hens	Average Production 1st year	2nd year	Average Production 1st year	No. of Hens 1st year	2nd year	Average Production 1st year	2nd year
All dams and daughters	77	223.76	155.42	220.78*	195	141	218.78	155.97
251- to 300- egg dams	13	263.92	175.16	220.78	50	41	229.26	163.61
201- to 250- egg dams	49	225.67	152.93	220.78	120	81	214.15	150.87
151- to 200- egg dams	15	182.73	143.7	220.78	25	19	220.08	161.21

*Yard average.

TABLE XLIII. SUMMARY OF FIRST YEAR'S PRODUCTION AND BEST TWO MONTHS, OREGONS, OREGON STATE HOSPITAL

Yard	Year	No. hens	First year	Best 2 months	Dam's average	Sire's dam's average
					Yd. E	
1	1914-15	33	216.15	48.88	(220.78)	201.00
2	1914-15	32	225.41	49.66	(220.78)	201.00
3	1914-15	32	222.38	49.72	(220.78)	201.00
4	1914-15	39	224.00	50.46	219.85	201.00
5	1914-15	36	214.44	51.56	221.19	201.00
6	1915-16	195	218.78	50.73	229.17	291.00

to lateness of hatch and other conditions. The high second-year record indicates high laying capacity. These pullets were the fourth generation from high-record pedigreed stock.

The June-hatched pullets of yards 4 and 5 were from the same yard E, but they had individual hen pedigrees. In other respects their pedigrees are the same as yards 1, 2, and 3.

In the winter of 1915 the pullets of yards 1, 2, and 3 were mated to males whose dam was C543, record 291 eggs. From this mating 195 pullets were produced and trap-nested. Their production is shown in Table XLI. They were from 77 of the hens in the yard, each chick being pedigreed. The dams averaged 223.76 eggs and the daughters averaged 218.78, while the dams' dams averaged 220.78. In the second year 141 of these pullets were kept over and averaged 155.97 in the second year, while their dams averaged 155.42.

In Table XLII have been grouped the 77 dams of 195 pullets of yard 6. It is seen that 13 of the dams laid an average of 263.92 eggs and within a range of 251 to 300 eggs inclusive; 49 averaged 225.67 and laid within 201 to 250, while 15 averaged 182.73 and laid within 151 to 200. Their second-year record is also given.

In the same table is shown the production of the 195 pullets for the first year, and the second-year production of 141 remaining. They are also grouped according to dams, the thirteen highest-record dams having 50 daughters with average record of 229.26 in first year. The second group of 49 dams produced 120 pullets with average record of 214.15, and the third group of 15 dams produced 25 pullets averaging 220.08 in first year.

There is no significant difference in the production of the pullets from the different groups of dams, the fifteen dams averaging 182 plus eggs producing pullets of but slightly lower average than the highest group of dams with average egg production of 263 plus, while the pullets from the second group of dams averaging 225 plus averaged a trifle less than the pullets from the lowest group of dams.

The dams, however, are practically all of high record and of the same breeding, and were mated to sires from the same dam. The records of the pullets in this case appear to follow the average production of the strain or of the ancestors of the female line. This result emphasizes the importance of the strain or of long high-production pedigree, rather than the value of the immediate parent.

It is rather remarkable how closely the records of the State Hospital flocks follow those of the Experiment Station flocks. Under different environment these flocks at the State Hospital showed practically the same production as the Station's flocks of similar breeding, indicating, as in the case of the Station flocks, transmission of high fecundity. The influence of the male whose dam laid 191 eggs, did not result in raising the production of the pullets above the average of the dams of four generations, the production of the three generations in the female line being strikingly similar. High fecundity, however, was maintained throughout the four generations. High fecundity in this case was represented by 200 eggs or more in a year, though the original high foundation stock laid less than half as many eggs.

In subsequent years, interruptions due to war service interfered with the keeping of many pedigrees, but the average of the whole flock, which now numbers about 4000 fowls, has been about the same in each year.

PRODUCTION PEDIGREE OF 195 PULLETS YARD 6. THE LOWER HALF
SHOWS PEDIGREE OF YARDS 1, 2, 3, 4, AND 5

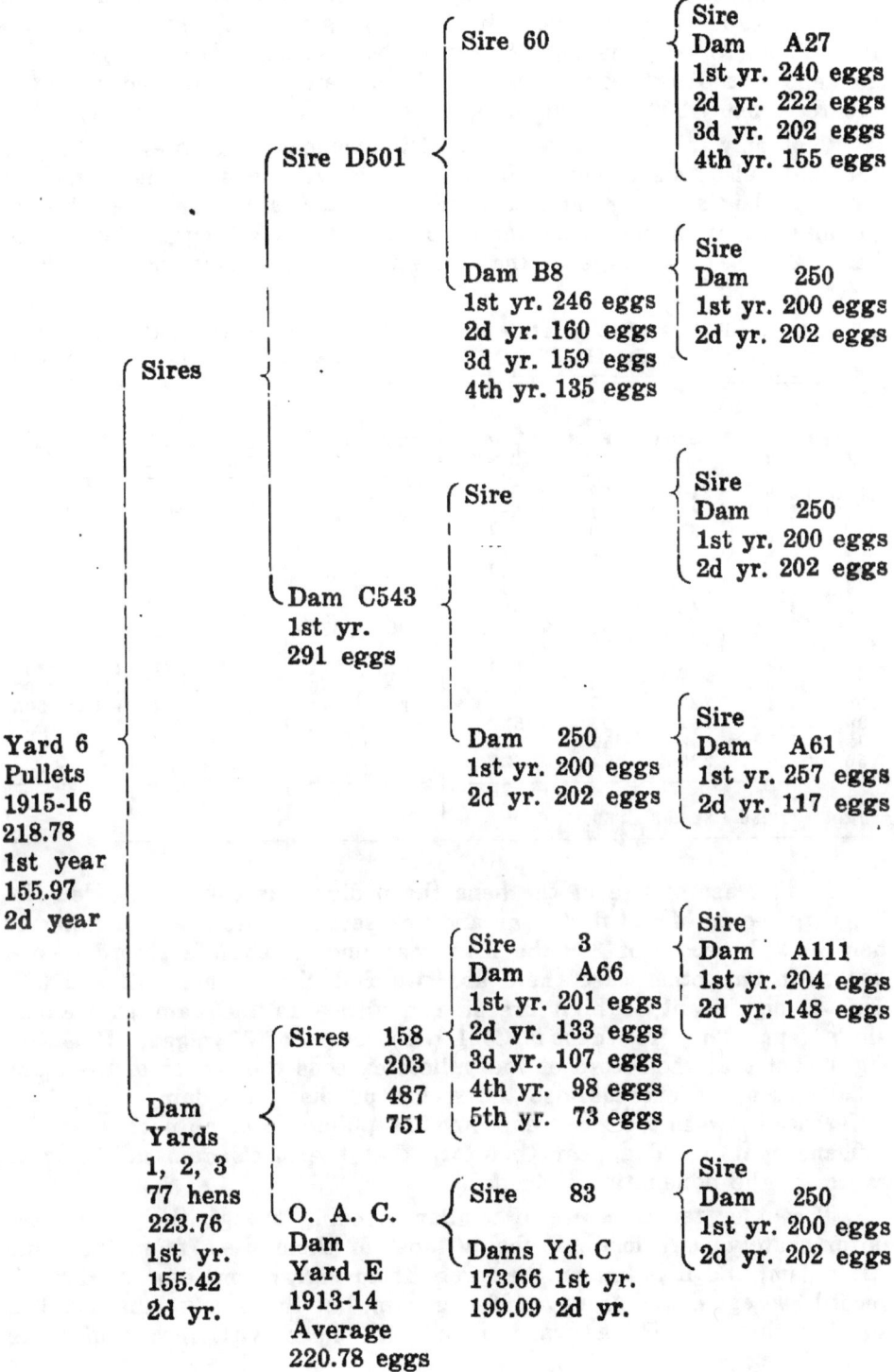

300-egg Hens. In 1918-19 a flock of about 430 pullets (the exact number cannot be given) were trap-nested at the State Hospital, and in the flock 15 hens made records in their laying year of 300 eggs or more (see Table XLIV), the highest record being 330. The average of the 15 hens was 310.86 eggs. It is no doubt an unprecedented record to secure so many 300-egg hens in one year from a flock of that size.

Not only is the first-year record phenomenal, but the second year record is also phenomenal. In the second year one of the hens died before it laid any eggs and another died after laying 152. All the remaining 13, with one exception, laid more than 200 eggs, the highest being 264. The average of the thirteen for the second year is 224.92.

TABLE XLIV. ANNUAL AND BEST TWO-MONTHS PRODUCTION OF 300-EGG HENS—OREGON STATE HOSPITAL

Hen No.	First year	Second year	Best 2 mos. First year	Total 2 years	Dam	1st yr. eggs	Dam's dam	1st yr. eggs	Sire's dam	1st yr. eggs
237....	330	216	59	546	....	......	......	......	......	......
29....	323	242	61	565	285	274	16	258	C 521	303
190....	323	231	60	554	....	......	......	......	......	......
264....	314	205	62	519	....	......	......	......	......	......
297....	313	264	58	577	....	......	......	......	......	......
444....	313	206	60	519	....	......	......	......	......	......
82....	310	243	59	553	....	......	......	......	......	......
38....	309	......	60	....	285	274	16	258	C 521	303
430....	308	257	59	565	....	......	......	......	......	......
72....	308	173	58	481	482	274	82	281	C 521	303
54....	306	-D	58	....	640	190	317	302	C 251	303
305....	304	223	58	527	482	274	82	281	C 521	303
396....	301	231	57	532	....	......	......	......	......	......
232....	301	208	59	509	....	......	......	......	......	......
156....	300	225	59	525	....	......	......	......	......	......
Average..	310.86	224.92	59.13							

In the case of five of the hens the pedigree is complete. Hen No. 29, with record of 323 first year and 242 second year, is a full sister of hen 38 with record of 309 the first year and incomplete record second year. In one other case there are two full sisters, hens 72 and 305. This is significant of high breeding qualities in the dam as well as sire's dam. The sire's dam is C521 with record of 303 eggs. It is also significant that the record of the pullets exceeds the record of the dams in all cases and exceeds also the record of the sire's dam, though the difference between the sire's dam and the pullets is of no particular significance. It would appear therefore that the sire's dam in this case exercised the dominating influence.

The other ten hens are of similar breeding though their pedigrees unfortunately were lost. Probably most of them were from the same sire's dam, and it is known they were from high-record dams. Hen 54, record 306 eggs, was from a 190-egg dam, but the dam's dam laid 302, so that this hen has a grandam on both sides with record of more than 300 eggs.

BREEDING QUALITIES OR PREPOTENCY VERSUS EGG-LAYING QUALITIES AS A BASIS OF SELECTION

When the term "pedigreed high producers" is used in this bulletin it is meant to convey the meaning merely that the hen or the male in question is from a high-record dam and sire's dam. It does not mean that their breeding qualities have been tested and found good. In point of fact in the majority of cases nothing was known of the ability of the breeding males or females to transmit their qualities. It would be difficult to prove whether the increased production was due to the effect of breeding the best producers to the best producers, or of selecting among the high producers, those that were good breeders of high producers.

In other words, was the increase due to selection on the basis of egg production only, or was it due to selecting for breeding stock only those individuals that had been proved to be capable of producing high producers?

In the first years of the experiment, there was no opportunity to select by the progeny test. The only basis of selection was the egg record, and it is true, from the records of the progeny, that this method of selection was effective in all three breeds. There are indications in the records that some hens produced better layers than others; in other words, while the difference in some hens does not show by their egg records, there does appear a difference in their breeding powers, judging by the records of their progeny. It was found that there were differences in males in transmitting their laying qualities, and this fact will be discussed later. This difference, however, is not so conspicuous as to warrant any deduction that the results were due in any large measure to differences in the breeding powers of males.

The point to be emphasized in this connection, however, is that a definite increase in production was secured by breeding the best to the best; that is, by using the annual record only as a basis of selection, in other words mating high-record hens to males from high-record hens, to produce high-record progeny.

It may be asserted that it "happened" that the breeding stock used possessed the powers of transmitting high production. If so, then it may be assumed, that this is just what is likely to "happen" in all flocks where numbers are used; and that regardless of the tested breeding powers of the breeding stock, or without regard to the record of their female progeny, there is bound to be increased production if the best is bred to the best, or if the breeders are selected only because they are the best, according to their own trap-nest records. This subject will be discussed in a later bulletin in studying the records of the progeny of different males.

VARIABILITY IN HIGH AND LOW PRODUCTION

Variability in production did not decrease with increased production. In other words, the range between the highest and lowest individual records was practically the same at the end of the experiments as at the beginning. This is shown in Figs. 11, 12, and 13 of frequency distribution of hens which give the number of hens each year in groups within a range of 20 eggs. For example, in the first year, 1908-09, with a total of 92 Barred Plymouth Rock hens, in the lowest group laying between 1 to 20 eggs, there was 1 hen; in the group laying between 21 and 40 there were 6 hens; in the next group laying between 41 and 60 there were 20 hens, and in the next, laying between 61 and 80, there were 25 hens. By looking to the right in the table of Barred Rocks under the year 1912-13 of pedigreed stock it is seen that, though there were more hens in the flock, there was only one hen in those four lowest groups.

In the first year the highest number of hens was found in the group laying between 61 and 80, while in the year 1912-13 of pedigreed stock the highest number was found in the two groups ranging from 161 to 200. In the last year the highest number of hens is found in the group laying 201 to 220 eggs. As production increased there were fewer hens in the lower groups. The average hen population moved up to higher levels as a result of better breeding.

With the Leghorns the mean moved up from group 101 to 120 to 201 to 220, while in the case of the Oregons the mean moved up from group 121 to 140 to group 221 to 240. The first year of the Oregons gives the record of the first crosses and not the foundation stock of Leghorns and Rocks of the previous year.

In the last year there was apparently as great an opportunity for selection as at the beginning of the experiment.

It will be noticed also that there was a movement upward among the high groups, almost arithmetical in proportion with the lower groups. Higher levels of production were reached. Low-producing hens were being eliminated and at the same time higher-record hens obtained.

This may not mean, however, that there will be relatively fewer slacker hens or unprofitable hens as the low-producing hens are eliminated. The costs of production determine whether a hen is a slacker or not. A hen that was profitable in 1908-09 might be a slacker in 1917-18 because in the latter year the costs might have increased to such an extent that her production was not profitable. Raising the costs increases the slacker population, while lowering the costs decreases the slackers. While our experiments had to do with production, not with costs, the practical effect of increased production is increased profits by elimination of slackers.

The same results are shown in the different breeds—a raising of the general average by breeding but practically the same variability in production at the beginning as at the end of the experiments.

The general average production of the hen population is shown by the heavy lines. Naturally the range will vary as the numbers in the flocks increase or decrease, but there is no mistaking the significance of these tables. While the average production of the progeny of selected high-producers is lower than the parents (dam and sire's dam) except in isolated cases already referred to, there was a progressive increase each year, individuals showing higher production than the parents in accordance with the principle or law of progression.

The actual high and low individual records of pedigreed parents and their pullets were as follows:

	1912-13		1913-14		1914-15		1915-16		1916-17		1917-18	
	High	Low	High	Low	High	Low	High	Low	High	Low	High	Low
B. P. Rocks—Parents	259	105	259	96	259	107	......	......	268	185	270	161
Pullets	268	3	268	86	246	108	......	......	220	154	308	54
W. Leghorns—Parents	240	215	......	......	267	162	......	......	302	139	302	114
Pullets	267	119	......	......	302	100	......	......	264	68	300	28
Oregons— Parents	257	113	226	133	303	119	303	267	......	......	303	156
Pullets	303	123	278	58	309	77	286	156	......	......	304	123

Barred Plymouth Rocks.

Eggs	1908-09	1909-10	1910-11	1912-13	1913-14	1914-15	1916-17	1917-18
301-306								1
281-300								1
251-280				1	1			4
241-260			1	1	4	4		5
221-240				7	9	8		10
201-220	1		9	14	14	7	4	19
181-200	1	1	6	26	15	16	5	10
161-180	1	3	9	26	24	11	1	1
141-160	3	2	6	15	16	14	1	3
121-140	8	8	7	12	8	9		3
101-120	10	8		4	8	2		1
81-100	16	3		1	4			2
61- 80	25	4	1					1
41- 60	20	1	1					1
21- 40	6							
1- 20	1		2	1				
Total Hens	92	28	42	108	103	71	11	62

Fig. 11.

Figs. 11, 12, and 13. These charts show how breeding (1) increases the number of high producers, but (2) does not decrease variability; (3) how it increases average flock production, though the progeny has lower production on the average than the parents of high records selected from previous generations.

The solid line represents the average production per hen of the progeny and the dotted line the average production of the parents, or the dam and sire's dam.

In a few cases hens of low production were used as breeders, but in those cases they were mated to a sire from a high-production dam. As the dotted line in the charts (see Figs. 11, 12, 13) shows the average of the parents was high there were few low-record hens used for breeding.

PREPOTENCY, AND POOR LAYERS FROM GOOD LAYERS

While high production records of the ancestors did insure, on the average, high-producing progeny, there are many exceptions to this. Poor layers are sometimes produced by good layers.

On the other hand, we have had some cases where good layers came from poor layers. The most conspicuous example of this was the progeny of a Barred Plymouth Rock male D461-2 (Table XLV) whose dam and dam's dam and sire's dam were poor layers as shown

White Leghorns.

Eggs	1908-09	1909-10	1910-11	1912-13	1914-15	1916-17	1917-18
301-302					1		
281-300							3
261-280				1	5	1	7
241-260				2	12	6	19
221-240		1	1	2	11	13	21
201-220			2	5	12	23	22
181-200	1			1	7	23	17
161-180	5	4	2	2	4	17	13
141-160	5	2	2		3	11	7
121-140	9	1	1		1	15	2
101-120	9	2	1	1	1	5	3
81-100	8	4	1		1	1	
61- 80	6	5				2	
41- 60	1						1
21- 40	3	1					1
1- 20	3	2					
Total Hens	50	21	10	14	58	117	116

Fig. 12. (See note for Fig. 11.)

on pedigree herewith. When he was mated to a hen with a record of 119 eggs, a pullet was produced that laid 309 eggs, the highest record at this Station. The combined record of the dam and sire's dam was 263, or 46 eggs less than the daughter.

This high capacity was clearly due to the influence of the sire, because when he was mated to various other hens, the daughters were good producers, 18 of them having complete annual records, averaging

221.11 eggs. Of these, 7 were crosses, or Oregons, though the 309-egg pullet had a White Leghorn dam. The 7 daughters averaged 270.28, the average of the dams was 202.28, of the dams' dams 209.

The eleven remaining daughters were from pure Barred Plymouth Rock dams and they averaged 189.82 with high record of 260. The average of the dams was 190.09 and dams' dams 164.18.

This male apparently was strongly prepotent. Though from a line of poor producers he carried the factor of high production. His female ancestors, as far back as of record, had not high productive powers, but

Oregons.

Eggs	1909-10	1910-11	1912-13	1913-14	1914-15	1915-16	1917-18
301-309			1		2		1
281-300			1		7	1	3
261-280			2	7	7	2	9
241-260		1	2	9	14	3	13
221-240			2	6	17	4	21
201-220	2	2	6	14	10	1	12
181-200	5	2	4	6	14	1	9
161-180	8	12		4	9		6
141-160	9	6	1	1	11	1	
121-140	16	5	1				1
101-120	15	4					
81-100	6	2		1			
61-80		2			1		
41-60				1			
21-40		1					
1-20	2						
Total Hens	63	37	20	49	92	13	75

Fig. 13. (See note for Fig. 11.)

through the male they transmitted high fecundity to the offspring. Unfortunately after his breeding characteristics had been discovered he would not breed, and there was no progeny except in this one generation.

It is not clear, so far as we have been able to analyze the records of this case, as to whether high prepotency appears in the later progeny of the first generation.

The daily egg record does not show high rate of laying in the dam of the 309-egg hen. The highest monthly production was 21 eggs in

May and 40 eggs as the best two-months production. The egg record of the dam of D461-2, also shows a low rate of laying, or at best a medium rate. The highest month was April, 24 eggs, and the best two months, April and May, was 43 eggs. In the second year her highest production was 22 in March. The dam's dam of the male with first-year record of 74, second year 91, third year 64, shows also low rate. In her first year, her best record was 20 eggs in March.

TABLE XLV. PEDIGREE OF BARRED ROCK MALE D 461-2 AND
PRODUCTION OF DAUGHTERS AND PEDIGREE OF DAMS

```
                                         ┌Sire
                          ┌Sire..........┤
                          │              │Dam 207
            ┌Sire 297.....┤              │ 1st yr. 169 eggs.
            │             │              │ 2d yr. 127 eggs.
            │             │Dam A 94      │ 3d yr. 113 eggs.
            │              1st yr. 20 eggs
D 461-2.....┤              2d yr. 64 eggs (D)
            │
            │             ┌Sire
            │Dam H 12 N...┤
             1st yr. 144 eggs.
             2d yr. 105 eggs. │Dam 66
             3d yr. 127 eggs    1st yr. 74 eggs.
                               2d yr. 91 eggs.
                               3d yr. 64 eggs.
                               4th yr. 64 eggs.
```

DAUGHTERS

No.	Production 1st yr.	Dam	Eggs 1st yr.	Dam's dam	Eggs 1st yr.
		Oregons			
E 138	309	C 584	119	O 34	229
E 178	299	C 503	207	A 47	205
E 156	284	C 518	226	A 47	205
E 127	274	C 503	207	A 47	205
E 114	262	C 470	211	A 47	205
E 147	241	C 503	207	A 47	205
1604	223	C 481	239	Yd. 1	
Average	270.28		202.28		209
		Barred Rocks			
E 659	260	C 117	186	A 79	219
E 719	219	C 76	198	B 154	191
E 652	192	C 153	177	66	74
E 669	190+	C 76	198	B 154	191
E 582	188	C 76	198	B 154	191
E 619	185	C 117	186	A 79	219
E 657	183	C 153	177	66	74
E 686	178+	C 76	198	B 154	191
E 678	178	C 76	198	B 154	191
E 709	174	C 76	198	B 154	191
E 587	141	C 153	177	66	74
Average	189.82		190.09		164.18
		Both Breeds			
Average	221.11		194.83		180

Neither by the annual record, nor by rate of laying is there any proof that the ancestors of this male, so far as we have trap-nest records of them, had high laying capacity. While the ancestors of high laying capacity produced, on the average, high laying progeny, and while the more immediate parents exercised the greatest influence; in other words, while high-record pedigreed breeding stock produced high-record progeny, there are instances such as this, where the production of the progeny cannot be accounted for by the pedigree of production of the ancestors, unless it can be explained on the assumption that in a more remote ancestor or ancestors, high fecundity or high prepotency may be lost apparently and recovered in subsequent generations.

While the breeder must put his reliance largely upon pedigree of production records in selecting his breeding stock, the result secured from this male shows that there is another factor to be reckoned with, which, in the absence of a more comprehensive definition, may be called prepotency. If he wishes to make the most certain progress in increasing production by breeding, he should not be content to breed the best layers to males of the best layers, but he will breed the best breeders to the best breeders, judging from the egg records of their progeny.

Our records point out two practical conclusions of first importance that must be drawn from these experiments; first, starting with an unselected flock of layers, and using each year only those hens with high trap-nest records and males from high trap-nested hens, there is certain to be an increase in production, always providing that environmental conditions are equally favorable each year; second, by selecting the breeding stock on the basis of the egg record of the progeny, the increase may be more certain and rapid. Such males, however, as D461-2 are rare, according to our records, and only by testing the progeny of a great many sires and dams each year would a breeder be likely to find phenomenal individuals.

The pedigree of this male is given in Table XLV, together with pedigree of his daughters and record of first year's production.

RELATION OF SHORT PERIOD PRODUCTION OR RATE OF LAYING TO ANNUAL AND BIENNIAL PRODUCTION

A study of our records shows that there is a pretty accurate correlation between monthly or bimonthly production and the annual record. If it should prove that a short-period record is as good a measure of the hen's inherited capacity as the annual record, then the problem is very much simplified.

Does the rate of laying or the intensity of egg production indicate egg-laying capacity? By rate of laying is meant the relative frequency that a hen lays in a short period. If the hen lays every day without missing a day, for five or six days or more, then misses one day and begins again on another long stretch, we say that she has a high rate of laying. Another hen that lays one day and misses one, and continues this way for an indefinite period, has a low rate of laying. Our records show characteristic differences among hens and the question is, can we predict with any degree of accuracy the probable annual production of a hen from a monthly or bimonthly study of her records. Environmental factors which can not very well be controlled may interfere with the hen's laying her full annual capacity. The longer the period of the record, the greater the likelihood that some disturbing factors will interfere with production.

The records are conclusive in showing that egg production may be increased by selective breeding, based on the annual trap-nest records. The increase is such that it could hardly be ascribed to possible changes in environmental conditions of different years. The same feeding system was used throughout all the years, with some minor changes in the ration. The same housing was used. The same size of yard and flocks was maintained. In fact the conditions, on the whole, were more favorable in the first year than in later years, so far as regards housing and soil. In the first year they were new and the yards were fresh and clean, though later differences were minimized as much as possible by rotation and cultivation of yards and by disinfection of the houses.

While the annual trap-nest record of the pullets is undoubtedly the best measure of a hen's laying capacity, under favorable conditions, it is not always the best, because of the difficulty of controlling the environment. If we select for breeding a hen that has made a record of 300 eggs a year, her record would indicate that her environment was favorable. If her daughters averaged 300 eggs we might fairly conclude that they inherited the capacity of the dam, and that they were kept under favorable environment. If, however, they averaged 150 eggs it might be that they inherited low production, but on the other hand, it might be that they had equal laying capacity with the dam, but through unfavorable conditions of environment their year's record did not represent their capacity.

In a shorter period, possibly, the real laying capacity of the hen may be better shown. An attempt has been made in Tables XLVI, XLVII and XLVIII to correlate the shorter-period records, in this case the best two-month records in the year, with the annual record. Spring is the season of maximum production. It is the natural laying season, and it is reasonable to expect that the full capacity of the hen can only be secured when the conditions for production are at their best. The question then is, does the hen that lays ten or twelve months of the year, lay more eggs than the hen that lays only in the spring months? Is it true that the hen that lays heaviest in the spring months does this because she started late and discontinued early? In other words, does a moderate layer do better in the spring than a good layer? Is it true that the hen that lays heavily in the spring does so because she has been reserving her energy for spring production, and is able to beat the long-distance rival in the short period?

BEST LAYERS BEST IN ALL MONTHS

The records show conclusively that the best layers throughout the year lay the largest number of eggs in the spring or any other season. The poor layer is relatively poor in any month of the year, though she lays a larger percentage of her eggs in the spring months than the good layer.

With a correlation established between the spring months or the best two-months production of the hen, and the annual record, we can predict with some accuracy from the best two-months production what the production should be for the year. The production for the best two months affords a good basis for judging the capacity for the hen.

TABLE XLVI. SUMMARY OF ANNUAL PRODUCTION OF HENS GROUPED ACCORDING TO BEST TWO-MONTHS RECORD
BARRED PLYMOUTH ROCKS

Yard	Year	No. hens	55 up		51-55		46-50		41-45		36-40		31-35		Below 31		Annual record by years*
			No.	Average	No.	Average	No.	Average	No.	Average	No.	Average	No.	Average	No.	Average	
4 & 5	1908-09	92	...	......	...	......	7	155.71	11	107.36	10	106.50	17	84.94	47	60.13	82.67
6	1909-10	28	...	......	...	......	...	......	5	148.40	6	136.83	8	112.50	9	92.33	117.64
15	1910-11	42	...	......	1	204.00	5	222.00	13	182.77	12	163.83	7	141.43	4	37.25	161.78
6	1912-13	38	1	196.00	6	218.50	11	187.82	13	171.38	3	144.80	1	89.00	1	3.00	174.52
7	1912-13	37	...	......	4	210.75	15	184.80	10	175.38	3	134.00	2	134.00	...	......	174.13
8	1912-13	33	1	196.00	5	200.60	8	185.50	12	158.50	7	164.57	1	139.00	1	107.00	
B	1913-14	52	5	219.80	16	213.13	18	185.28	9	143.58	3	141.67	3	134.00	...	......	180.48
C	1913-14	51	2	196.50	11	191.55	16	166.75	9	160.56	5	138.60	3	126.00	2	125.50	
L	1914-15	33	5	215.20	3	182.00	12	176.42	13	177.67	5	143.40	2	124.00	...	......	194.45
P	1914-15	38	2	155.00	6	227.17	9	193.00	9	155.08	4	166.75	1	135.00	...	......	
18	1916-17	11	...	......	4	187.00	3	200.67	2	199.50	...	......	...	......	...	......	202.40
E	1917-18	62	4	281.00	17	229.41	21	207.90	14	185.57	...	......	4	103.25	3	148	
Total..........		517	20	4473	73	15435	125	23547	124	20127	60	8632	49	5406	66	4315	
Average.....				223.55		211.44		188.38		162.68		143.88		110.33		65.38	158.57

*Unidentified eggs not counted.

TABLE XLVII. SUMMARY OF ANNUAL PRODUCTION OF HENS GROUPED ACCORDING TO BEST TWO-MONTHS RECORD
WHITE LEGHORNS

Yard	Year	No. hens	55 up		51-55		46-50		41-45		36-40		31-35		Below 31		Annual record by years*
			No.	Average	No.	Average	No.	Average	No.	Average	No.	Average	No.	Average	No.	Average	
4 & 5	1908-09	50	...	......	...	......	1	177.00	10	154.60	12	122.42	11	107.27	16	58.44	106.14
9	1909-10	21	...	......	...	......	2	203.50	3	151.33	4	139.25	4	75.00	8	60.00	104.67
1	1910-11	10	...	......	1	240.00	3	200.00	2	162.50	2	130.00	1	98.00	1	123.00	164.60
3	1912-13	14	1	283.00	6	234.17	4	209.75	2	192.00	2	141.00	...	......	...	......	207.85
3	1914-15	16	3	254.00	3	237.00	10	207.20	...	......	...	......	1	100.00	...	......	216.76
O	1914-15	42	...	......	14	237.71	17	206.76	7	176.43	5	159.20	...	......	...	......	
P	1916-17	49	3	207.00	13	211.77	24	188.25	7	160.43	7	129.00	4	96.75	1	95.00	181.50
15	1916-17	57	2	233.00	6	221.50	22	187.09	15	170.47	...	......	...	......	...	......	
F	1917-18	11	1	280.25	6	221.17	2	206.00	1	160.00	4	150.50	...	......	1	23.00	209.24
G	1917-18	60	4	258.50	12	239.33	30	252.20	9	191.00	2	137.00	1	56.00	...	......	
		56	2		20	211.60	20	209.65	11	174.45							
Total..........		386	16	4129	81	18077	135	27593	67	11422	38	5143	22	2121	27	1656	
Average.....				258.06		223.17		204.39		170.48		135.34		96.41		61.33	181.71

*Unidentified eggs not added.

TABLE XLVIII. SUMMARY OF ANNUAL PRODUCTION OF HENS GROUPED ACCORDING TO BEST TWO-MONTHS RECORD OREGONS

Yard	Year	No. hens	55 up No.	55 up Average	51-55 No.	51-55 Average	46-50 No.	46-50 Average	41-45 No.	41-45 Average	36-40 No.	36-40 Average	31-35 No.	31-35 Average	Below 31 No.	Below 31 Average	Annual record by years*
7..	1909-10	33	....	....	....	....	1	186.00	2	200.00	5	154.40	16	134.31	10	96.70	133.76
8..	1909-10	30	....	....	....	....	....	....	3	184.67	7	159.00	9	133.33	10	108.60	
1..	1910-11	37	....	....	....	....	5	189.20	8	170.75	9	162.67	7	139.71	8	86.37	147.16
9..	1912-13	20	4	278.25	3	245.33	9	208.67	2	196.50	1	146.00	....	....	1	123.00	219.45
E.	1913-14	49	7	233.86	24	227.67	11	212.27	4	179.00	3	151.67	....	....	....	....	216.74
4..	1914-15	10	2	266.50	4	253.25	3	236.67	1	204.00	....	....	....	....	....	....	217.24
5..	1914-15	10	1	200.00	4	188.00	4	216.50	1	152.00	....	....	....	....	....	....	
6..	1914-15	14	2	269.50	6	221.33	5	182.40	1	177.00	....	....	....	....	....	....	
J..	1914-15	58	11	276.09	16	227.87	19	209.11	7	169.86	3	176.33	1	149.00	1	77.00	
10.	1915-16	13	2	278.50	5	238.40	4	227.25	1	233.00	1	156.00	....	....	....	....	234.39
7..	1917-18	12	1	270.00	5	242.60	4	249.25	2	220.00	....	....	....	....	....	....	227.42
H.	1917-18	63	12	244.08	32	230.56	16	208.06	2	188.50	1	123.00	....	....	....	....	
Totals....		349	42	10815	99	22722	81	17041	34	6201	30	4758	33	4476	30	?944	
Average..				257.50		229.52		210.38		182.38		158.60		135.64		98.13	197.58

* Unidentified eggs not added.

It may happen, however, and it is shown in our records, that a hen with high spring production does not make a high annual record due to unfavorable conditions. These conditions may be weather changes, mistakes in feeding or care, diseases, broodiness, lack of vigor, and other things. Such a hen, judged by the annual record is classed as a poor layer, when in point of fact, she has high egg-laying capacity, judged by a shorter period, and in breeding may be as valuable as the hen with the high annual record, as a breeder producing possibly as good layers as the high annual-record hen.

This may explain in part, why good layers are frequently produced by hens with low annual records.

The best two months are usually March and April, and April and May, though other months occasionally show the highest records.

VIGOR

Another factor that comes into the problem is that of vigor. It is assumed that the hen must have good vigor to stand the strain of heavy production for a year, so that any method of breeding that increases vigor should increase production. It might be a question then whether the increase in production secured by breeding from selected high producers did not primarily come from inherited vigor. That is another problem.

It has been shown by Doctor Pearl (Maine bulletin 205, Nov. 1912) that poor layers as well as good layers have more eggs or oocytes in the ovary than they have ever been known to lay, and that poor layers inherited the eggs or oocytes apparently in as great numbers as the good layers so that the inheritance of high fecundity is not a question of inheriting a great number of eggs, but rather a question of inherited capacity to lay them.

Our records show a large increase in production by breeding from high producers, and that the increase came by inheritance, but it would appear that it was rather ability to lay that was inherited than that eggs were inherited. If this ability to lay means greater vigor, is it possible to measure laying vigor by a study of our records? Vigor or lack of vigor is often evident to the eye, but not always. Very frequently the hen having all the evidences of good health and vigor does not show it in her egg record. If long-continued production indicates vigor, then it is easy to select for vigor by selecting hens with the longest productive period. Our records, however, show in many cases long production with medium low records. In these cases the records show that the hen would lay one day and miss the next or lay two days in succession and miss the third. In other words, the rate of laying was low. On the other hand, a high producer misses fewer days in the month and may lay four or five days in succession, and even many days before missing. That is a high rate of laying or intense production. Both hens are comparatively long layers. The records show another class of layers, the very poorest. This class shows a low rate of laying, and short-period production. In the first case the hen has vigor, but has not the inherited ability to lay many eggs, whatever that ability may be. But with the vigor she makes a medium good layer or medium poor layer. In the second case the hen has both high vigor and inherited ability to lay, as shown by her high rate of laying, as well as long period of high production. In the last case the hen has low vigor as well as low ability to lay, as shown by her low rate of laying and her short-period production.

Four Types of Layers. Laying hens may be divided into four distinct types, so far as their yearly trap-nest records show, as follows:

(1) The daily record of the hen of the first type shows that she starts early in the year and continues late at the latter part of the laying year. Her rate of laying is high. That is, she will miss comparatively few days. She may lay four or five or more days and miss one day, and may lay sometimes a week or two weeks without a break, and her annual record will be 200 eggs or more, depending upon how many days in succession she usually lays. This hen has high fecundity plus high vigor.

(2) The second type of hen may lay 150 eggs distributed over a comparatively long period, but her rate of laying is low. She may lay one or two days and miss one and continue in that way. She has not high capacity as shown by her low rate of production and her low best two months, but she has vigor, as shown by her steady or consistent laying. This hen has low fecundity plus vigor.

(3) The third type of hen shows a high rate of laying in the best two months but in the later months of the year she shows either a diminished rate of laying or she discontinues altogether. She may lay 150 or she may not lay 100. This hen has inherited high fecundity but lacks vigor.

(4) The fourth type of hen begins to lay late, possibly two months after the first hen started, and stops early in the year. Her laying year is short at both ends, and her rate of laying is also low. Her best two-months record is low. She has poor laying capacity and low vigor.

The reproduction in Figs. 14 to 17 of the actual daily records of hens illustrates the four different types.

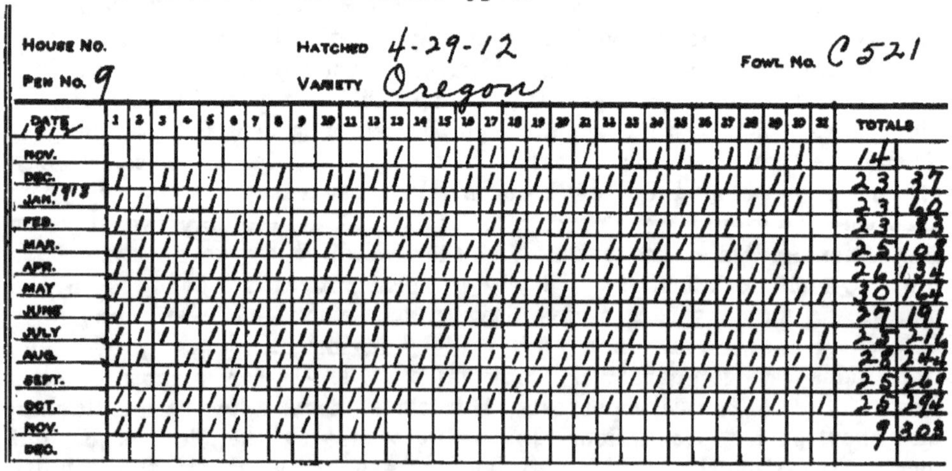

Fig. 14. Daily record of C521 showing high rate of laying through twelve months, indicating (1) inherited high laying capacity, (2) good vigor, and (3) favorable environment.

HOUSE No. *A*
PEN No.
HATCHED 2-13-11
VARIETY White Leghorn
FOWL No. B15 / H ♂ A

DATE	1	2	3	4	5	6	7	8	9	10	11	12	13	14	15	16	17	18	19	20	21	22	23	24	25	26	27	28	29	30	31	TOTALS	
1911 SEPT.																															/		
OCT.																															/		
NOV.		/	/			/	/												/		/		/	/		/	/					14	15
DEC.	/		/				/	/			/									/												7	22
JAN. 1912		/				/	/	/												/												15	37
FEB.																																15	52
MAR.	/	/	/		/	/																										18	70
APR.	/	/	/		/																											17	87
MAY	/	/																														16	103
JUNE	/	/																														16	119
JULY		/				/														/		/		/	/		/	/				15	134
AUG.		/	/		/																		/			/		/				14	148
SEPT.												/	/		/	/		/	/		/	/		/	/							10	158
OCT.	/			D																												1	159

Fig. 15. Daily record of B15 showing low rate, but consistent laying throughout the year, indicating (1) poor laying capacity, (2) good vigor, (3) favorable environment. Note Figs. 18 and 19, giving record of sister and dam indicating inherited low rate of laying.

HOUSE No.
PEN No. *C*
HATCHED 3-29-13
VARIETY Barred Rock
FOWL No. D181 / 583

DATE	1	2	3	4	5	6	7	8	9	10	11	12	13	14	15	16	17	18	19	20	21	22	23	24	25	26	27	28	29	30	31	TOTALS	
1913 AUG.																																	
SEPT.																																	
OCT.																																	
NOV.																																	
DEC.																																9	
JAN. 1914	/						/	/		/	/			/	/			/			/	/			/						/	18	27
FEB.	/	/			/	/	/																									22	49
MAR.	/	/	/	/	/	/	/	/																							/	25	74
APR.	/	/	/	/	/	/	/																									18	92
MAY	/	/			/	/																							X	/		16	108
JUNE			/	/			X																X							X		10	118
JULY					X																									X		13	131
AUG.	/	/	X																													10	141
SEPT.	D																																
OCT.																																	

Fig. 16. Daily record of D181 showing high rate of laying for short period, indicating (1) inherited good laying capacity; (2) poor vigor.

HOUSE No. *8*
PEN No.
HATCHED 5-3-19
VARIETY White Leghorn
FOWL No. J765 / 3126 J

DATE	1	2	3	4	5	6	7	8	9	10	11	12	13	14	15	16	17	18	19	20	21	22	23	24	25	26	27	28	29	30	31	TOTALS	
SEPT.																																	
OCT.																																	
NOV.																																	
DEC.																																	
JAN. 1920																																	
FEB.																	/	/		/	/		/	/		/	/					8	
MAR.	/	/	/	/	/																											13	21
APR.	/	/																														18	39
MAY	/		/	/		/	/																									15	54
JUNE																	/															8	62
JULY																		/														12	74
AUG.		/					/																									9	83
SEPT.																																	
OCT.																																	

Fig. 17. Daily record of J765, showing low rate, late laying maturity, inconsistent laying, early quitting at end of year, indicating poor laying capacity and poor vigor.

RATE OF LAYING INHERITED

Evidence that rate of laying is inherited is furnished by the daily record of B10, sister of B15, both showing same characteristic laying as their dam 47. The records of B10 and 47 are shown in Figs. 18 and 19.

Fig. 20 shows record of hen of very high capacity inherited from ancestors, but through a break in her record due to some unfavorable environment, including molting, her record does not indicate her full capacity.

HOUSE NO. *A*   PEN NO.   HATCHED 2-13-11   VARIETY W. L.   FOWL NO. *B10* *H7A*

DATE	TOTALS
1911 SEPT.	
OCT.	1
NOV.	12 / 13
DEC.	9 / 22
1912 JAN.	12 / 34
FEB.	15 / 49
MAR.	15 / 64
APR.	15 / 79
MAY	15 / 94
JUNE	12 / 106
JULY	10 / 116
AUG.	12 / 128
SEPT.	12 / 140
OCT.	5 / 145

Fig. 18. Daily record of B10, almost a complete duplicate of the record of her full sister B15. Poor laying capacity, but good vigor.

HOUSE NO. 9   PEN NO.   HATCHED 1909   VARIETY White Leghorn   FOWL NO. 47

DATE	TOTALS
1910 APR.	9
MAY	14 / 23
JUNE	14 / 37
JULY	4 / 41
AUG.	7 / 48
SEPT.	10 / 58
OCT.	12 / 70
NOV.	4 / 74
DEC.	6 / 80
1911 JAN.	18 / 98
FEB.	18 / 116
MAR.	18 / 134
APR.	6 / 140
MAY	

Fig. 19. Daily record of 47, showing same characteristic rate of laying as that of her daughters B10 and B15, indicating transmission of low rate of laying, and good vigor.

BEST TWO-MONTHS PRODUCTION AND LAYING CAPACITY

In Table XXXVI, Barred Plymouth Rocks, it is shown that from the first year to the last of the experiment there is a decided increase in the best two-months production. In other words the rate of laying, as well as the annual production, has been increased, though it will be noticed that on a percentage basis the rate of laying did not increase as much as the annual production. The fact that the best two-months production and the rate of laying increased on the average is corroborative evidence when taken in connection with the increase shown by the annual production, that the increase came by inheritance. The other fact that the best two-months production did not increase as much as the

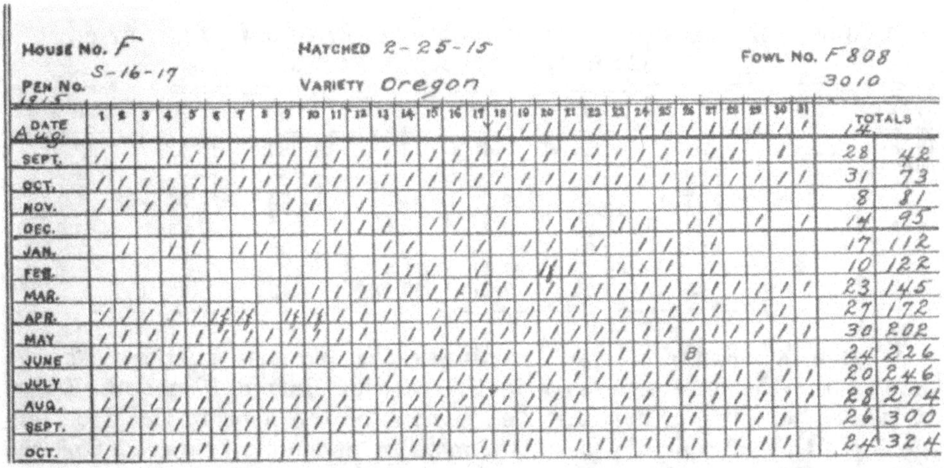

Fig. 20. Daily record of F808, granddaughter of C521, showing high rate of laying, but unfavorable environmental conditions part of the year. Her high egg-laying capacity is indicated (1) by her early laying maturity, (2) high rate of laying, (3) heavy laying at the end of the year. Her rate of laying would indicate that this hen had a capacity of about 300 eggs. Her record was 262 in the first twelve months of laying. Her sire's dam laid 291, dam 254, dam's dam 303.

annual record, would appear to be strong evidence that the environmental conditions under which the flocks were kept for the different years were fairly well controlled. Table LXI shows that on a percentage basis there is less difference between good and poor layers in the spring months than in other months. If the percentage increase in production showed as high for the best two months as for the year, it would then appear that the environmental conditions were not under proper control, and were such that the flocks did not lay throughout the year approximately their full capacity.

On the other hand, when the annual production is high, and the maximum bimonthly production, though high, shows a smaller percentage increase from low to high month than in cases where the annual production is low, it would appear evident that the conditions were favorable throughout the year.

It will be noted from the summary of the Barred Plymouth Rock for different years (Table XXXVI) that the average of the best two-months production of all years and flocks is 42.25 eggs a hen. In the first year the average of the 92 hens was 31.29 eggs a hen. In the last year the average of 62 hens was 47.19 eggs a hen. In the first year the annual production was 86.14 and in the last year 214.63. The increase in the annual flock record was 149.16 percent and in the best two months the increase was 50.8 percent. If we take the second year 1909-10 as a starting point, the increase in annual production is 77.85 percent, while the best two-months increase is 39.08 percent.

The best two months are, on the average, March and April. In case of individual hens, however, the best two months may be April and May, or May and June. The average for all the flock for best two months, as shown by the table, is 42.25 eggs a hen, while the average for March and April is 38.15 eggs a hen.

TABLE XLIX. PERCENTAGES BEST TWO-MONTHS PRODUCTION BARRED PLYMOUTH ROCKS

Year	Above 55	51-55	46-50	41-45	36-40	31-35	Below 31
1908-09	....	....	7.6	11.96	10.87	18.48	51.09
1909-10	....	....		17.86	21.43	28.57	32.14
1910-11	....	2.38	11.9	30.95	28.57	16.67	9.52
1912-13	1.85	13.88	31.48	33.33	13.88	3.7	1.85
1913-14	4.85	26.21	33.01	20 39	7.77	5.83	1.94
1914-15	9.86	12.68	29.58	30.99	14.08	2.81	....
1916-17	18.18	36.36	27.27	18.18	....	....	....
1917-18	6.45	27.42	33.87	22.58	6.45	3.23	....

Table XLIX shows by percentages the increase in the best two months for the different years in the case of Barred Plymouth Rocks. It is seen that in the first two years, none of the hens in the flock laid more than 50 eggs in the best two months, and only 7.6 percent of the flocks of 1908-09 laid more than 45 eggs, while better than 51 percent of them laid fewer than 30 eggs in the best two months. In the first year of the pedigreed stock 1.85 percent of them laid more than 55 eggs in the best two months, 13.88 percent laid between 50 and 55 eggs, 31.48 percent laid between 45 and 50 eggs, and only 1.85 percent of the flock laid fewer than 30 eggs in the best two months. There was the same percentage above 55 as below 30 in the last year's flock. In the rest of the years there is an increasing number in the high columns except in the last year, 1917-18. The explanation for this latter fact possibly is because this flock may have been in some respects under slightly more favorable conditions than those of the previous years. If conditions are favorable throughout the year there might be a slight decrease in the high months, if in the low months the production was unusually high. This was true of the 1917-18 flock. Their production in the low months was higher than that of the flocks of previous years.

If a good layer is not pushed, or for some other reason she does not lay her full capacity in the winter months, the presumption is that if she is in good physical condition she will lay a greater number of eggs in the spring months than she would otherwise. This does not alter the fact, however, that the poor or unselected layers in the first two years did not lay heavily in the season of maximum production, even though their winter production was also low.

There can be no other explanation of the showing made from the Barred Plymouth Rocks in this table, in connection with the record of annual production, except the fact of transmission of high fecundity.

In the case of the White Leghorns, the rate of laying is represented by 44.76 eggs a hen for the best two months for all the years, or 32.54 in the first year, and 48.16 in the last year, an increase of 48 percent, while the increase in the annual production is 98.72 percent.

If the hens are placed in different groups, according to the number of eggs they lay in the best two months, we find that there is, as in the case of the Barred Plymouth Rocks, a steady increase in the number of hens in the high groups. In the first two years of unselected or non-pedigreed stock, none of the hens laid more than fifty eggs in the best two months. Only 2 percent of them laid 46 to 50 eggs, inclusive, while 32 percent of them laid 30 eggs or less. In the first year of the pedigreed

TABLE L. PERCENTAGES BEST TWO-MONTHS PRODUCTION
WHITE LEGHORNS

Year	Above 55	51-55	46-50	41-45	36-40	31-35	Below 31
1908-09	....	....	2.00	20.00	24.00	22.00	32.00
1909-10	....	....	9.52	14.29	19.05	19.05	38.09
1910-11	....	10.00	30.00	20.00	20.00	10.00	10.00
1912-13	....	42.86	28.57	14.28	14.28	....	....
1914-15	10.34	29.31	46.55	12.07	....	1.72	....
1916-17	3.42	21.37	41.02	19.66	10.26	3.42	0.85
1917-18	5.17	27.59	43.10	17.24	5.17	0.86	0.86

TABLE LI. PERCENTAGES BEST TWO-MONTHS PRODUCTION
OREGONS

Year	Above 55	51-55	46-50	41-45	36-40	31-35	Below 31
1909-10	....	....	1.59	7.94	19.05	39.68	31.75
1910-11	....	....	13.51	21.62	24.32	18.92	21.62
1912-13	20.00	15.00	45.00	10.00	5.00	....	5.00
1913-14	14.29	48.98	22.45	8.16	6.12	....	....
1914-15	17.39	32.61	33.69	10.87	3.26	1.09	1.09
1915-16	15.38	38.46	30.77	7.69	7.69	....	....
1917-18	17.33	49.33	26.67	5.33	1.33	....	....

stock, 1912-13, 42.86 percent laid 51 to 55 eggs inclusive, and none laid fewer than 36 eggs. In the last three years a fair percentage of them laid above 55 eggs. Practically none of the pedigreed stock in all the years laid fewer than 31 eggs, while a large percentage of them in the unselected flock laid fewer than 31.

In rate of laying the Oregons show an increase in best two months between the first cross and the last of 58.05 percent.

In the percentage table (LI) where the flocks of each year are grouped according to best two-months production, we find that there were no hens in the flock of non-pedigreed stock, that laid more than 50 eggs in the best two months during the first two years. Only 1.59 percent of the first generation of crosses laid more than 45 in the best two months while 39.68 percent laid between 31 and 35 inclusive and 31.75 percent laid 30 eggs or less. Practically none of the pedigreed stock laid fewer than 36 eggs as their best two-months production, while in the first year of pedigreed stock, 20 percent of the flock laid more than 55 eggs and 15 percent laid between 51 and 55 eggs inclusive. In the last year more than 60 percent of the flock laid above 50 eggs a hen in the best two months.

The same general result was secured with Barred Plymouth Rocks, White Leghorns, and Oregons; namely, an increase in the egg-laying capacity of the hen, as judged by her record in the season of maximum production when the environmental factors are most favorable for determining her full capacity. It would appear that the characteristic of high rate of laying is inherited along with high annual production.

Culling Practice. An accurate culling practice may be based upon these findings. A low record for two months shows that the hen has low capacity and will make a low annual record. If the poultryman trap-nests his flock, he may discard or market all hens that do not lay more than 35 eggs in the best two months.

Table LII has been prepared to show what would be the result if from the 517 Barred Rock hens averaging 158.57 eggs a year we had culled out all hens that laid fewer than 36 eggs in the best two-months production. Of the total, 402 laid more than 35 eggs a hen in the best two months and averaged 179.75 eggs a hen. The remainder, or 115, laid 35 eggs or less in the best two months, with an average annual record of 84.53 eggs a hen. Among these, however, there were 7 hens that laid more than the average annual production of those in the higher group that laid 36 to 40 eggs. In other words, in culling out the 115 hens there was an accuracy of 93.91 percent. This is assuming that the hen of poor laying capacity is indicated by production of 35 eggs or less in the best two months.

If the White Leghorns were culled in the same way there is an accuracy of 97.96 percent. In other words, there were 49 culls in 386 hens, and 1 of the 49 laying fewer than 36 eggs in the best two months laid more in the year than the average of the 36 to 40 group. The 49 culls averaged for the year 77.08 eggs.

TABLE LII. PRODUCTION OF HENS LAYING 36 EGGS OR MORE AND 35 EGGS OR LESS IN BEST TWO MONTHS. BARRED PLYMOUTH ROCKS

Yard	Year	No. hens above 35	Annual Average	Hens laying less than 36	Hens laying less than 36 but more than annual production of group 36-40		Per cent Accuracy	Average production of culls	Above 36 but below 143.88 annual average
					No.	Annual Average			
4 and 5	1908-09	28	119.14	64	1	153.00	98.44	66.72	22
6	1909-10	11	142.09	17	1	179.00	94.12	101.82	5
15	1910-11	31	182.45	11	2	166.50	81.82	103.55	2
6	1912-13	36	181.25	2	0	....	100.00	46.00	7
7	1912-13	35	179.91	2	1	149.00	50.00	134.00	4
8	1912-13	31	174.84	2	0	....	100.00	123.00	5
B	1913-14	49	181.49	3	2	158.00	66.67	134.00	6
C	1913-14	46	174.15	5	0	....	100.00	125.80	9
L	1914-15	31	179.74	2	0	....	100.00	124.00	5
P	1914-15	37	185.38	1	0	....	100.00	135.00	7
18	1916-17	11	194.45	0	0	....	100.00	....	0
E	1917-18	56	214.07	6	0	....	100.00	93.50	2
Total....		402	179.75	115	7	161.43	93.91	84.53	74

With the Oregons the percentage of accuracy was 92.06. In 349 hens, 63 were culled out; and of these 5 averaged 165.4 eggs, which was more than the average of the group laying 36 to 40 eggs. The average of the culls was 117.78 eggs.

Of the total of the three breeds, 1252 hens, there were 227 culls and only 13 of them laid above the average of the higher group.

If culling is done on the basis of egg-laying capacity and if egg-laying capacity is accurately determined by rate of laying, or the best two-months production, an accurate system of culling is indicated by the results of these experiments.

TABLE LIII. PRODUCTION OF HENS LAYING 36 EGGS OR MORE AND 35 EGGS OR LESS IN BEST TWO MONTHS. WHITE LEGHORNS.

Yard	Year	No. hens above 35	Annual Average	Hens laying less than 36	Hens laying less than 36 but more than annual production of group 36-40		Per cent Accuracy	Average production of culls	Above 36 but below 135.34 annual average
					No.	Annual Average			
4 and 5	1908-09	23	138.78	27	0	....	....	78.33	8
9	1909-10	9	157.56	12	0	....	....	65.00	2
1	1910-11	8	178.12	2	0	....	....	110.50	2
9	1912-13	14	207.86	0	0	....	....	....	1
3	1914-15	16	227.00	0	0	....	....	....	0
O	1914-15	41	215.61	1	0	....	....	100.00	2
P	1916-17	49	187.55	0	0	....	....	....	4
Q	1916-17	52	176.90	5	1	136.00	80.00	96.40	11
15	1916-17	11	215.00	0	0	....	....	....	0
F	1917-18	59	221.53	1	0	....	....	23.00	1
G	1917-18	55	202.24	1	0	....	....	56.00	2
Total Averages		337	196.93	49	1	136.00	97.96	77.08	33

TABLE LIV. PRODUCTION OF HENS LAYING 36 EGGS OR MORE AND 35 EGGS OR LESS IN BEST TWO MONTHS. OREGONS.

Yard	Year	No. hens above 35	Annual Average	Hens laying less than 36	Hens laying less than 36 but more than annual production of group 36-40		Per cent Accuracy	Average production of culls	Above 36 but below 158.60 group average
					No.	Annual Average			
7	1909-10	7	167.43	26	2	163.50	92.32	119.84	3
8	1909-10	11	168.45	19	1	179.00	94.74	120.32	3
1	1910-11	22	171.64	15	2	160.50	86.67	111.27	6
9	1912-13	19	224.53	1	0	....	100.00	123.00	1
E	1913-14	49	216.47	0	0	....	100.00	....	3
4	1914-15	10	246.00	0	0	....	100.00	....	0
5	1914-15	10	197.00	0	0	....	100.00	....	1
6	1914-15	14	211.14	0	0	....	100.00	....	3
J	1914-15	56	220.96	2	0	....	100.00	113.00	5
10	1915-16	13	234.38	0	0	....	100.00	....	1
7	1917-18	12	243.33	0	0	....	100.00	....	0
H	1917-18	63	224.38	0	0	....	100.00	....	1
Total Averages		286	215.16	63	5	165.40	92.06	117.78	27

This does not mean, however, that these 227 culls represent all the hens in the flock that made poor annual records. We have found that there is a strong correlation between the best two-months production and the annual record, but that there are exceptions to this rule. There are individual hens with high rate of laying that make a more or less poor annual record, and the probable explanation is that environmental conditions in some way interfered with their laying their full capacity during the year. In such cases the hen probably has good laying capacity but does not show it in an annual record, and she may be a good breeder of good producers.

To show, however, how many hens were left in the flock laying more than 35 eggs in the best two months, but less than the average annual production of the next higher group laying 36 to 40 eggs, a count has been made. This shows that there were 74 of these among the Barred Rocks, 33 in the Leghorns, and 27 in the Oregons, a total of 134.

If the flocks were culled by the annual trap-nest record of each hen, there would have been culled out at the end of the year 348 hens, and of these 134 were probably of good laying capacity. That is, of the 348 culled out, 134 made a low annual record, but had good laying capacity, judged by their best two months.

If, however, the flocks were trap-nested only in the season of maximum production, and all that showed a low rate of laying were culled out, according to these results the culling would show an accuracy of over 90 percent. Practically all of the hens of poor laying capacity, judged by low rate of laying, would be culled out.

Conclusions. Intensity of production, or high rate of laying, in any two months of the year is a fairly accurate measure of the hen's laying capacity, and is a valuable check on the annual record in studying the effect of selective breeding.

The best two-months record as a means of predicting what the annual record will be is accurate within a certain range. Our records show many hens with high best-two-months production, but with low annual production. This is due to other factors and not to inherited laying capacity. The hen may inherit high laying capacity or high fecundity, but other factors will prevent her making a high annual record, though in a short period or in her best two months, she may make a high record, indicating high fecundity.

The hen may inherit high fecundity, and may make a high record part of the year, including a high March and April record, but she contracts disease that reduces her production, so that her annual record does not represent her full inherited capacity. She may go broody or may go into a moult, preventing her making a high annual record, though short period production may be high. She may not inherit high vigor and at the latter part of the year break under the strain of high production, though her best two-months production may be high and represent her inherited laying capacity better than her annual record.

The writer believes that this distinction between vigor and inherited egg-laying capacity should be made in studying breeding records. It should also be considered in selecting breeding stock for breeding experiments. It may happen that in breeding from and for low producers the progeny are high producers, because the selection was made by the annual record instead of the short-period production. In selecting breeding stock, as good progeny may be secured from a hen with a record of 150 eggs as from one of 250 eggs, if in each case they show equally high production in the high production period.

The annual record of the hen does not indicate always her inherited laying capacity, and therefore, does not always indicate her breeding capacity. A hen with a poor annual record may have inherited high capacity and may be a good breeder of high producers, and the short-period production is the best explanation of this fact.

RATE OF LAYING AND CHEMICAL COMPOSITION OF THE EGG
(Preliminary Report)

In so far as ordinary food analyses are concerned, hen's eggs are of strikingly uniform quality. It might be expected, however, that the individual variations which have been shown to be extremely great in number of eggs laid, would extend in some way or in some measure to the quality of the eggs. It is known that there are some differences in the keeping quality of the eggs and considerable differences in the hatching quality as well as in quality or vigor in the chicks hatched.

The investigations reported herewith had to do with the relation between rate of laying, or the intensity of production, and the fat content of the eggs. Some important preliminary work was done in 1920 by Mary Vernon Skelton as a basis for thesis work for a Master's degree in Home Economics, on "The Fat Content of Eggs of Good and Poor Layers." The question, however, of whether the hens had high or low annual record was not involved. A good layer in this case was one with high rate of laying, as shown by her egg record for several weeks preceding the collection of eggs for analyses, and the poor layer was one of comparatively poor rate of laying. In point of fact all the hens used in this experiment were of high annual record. They were all hens that had concluded at least one year's laying and some of them several years. A record of their production for eight weeks ending April 17, 1920, together with total number of eggs laid in their first or pullet year is shown below. The eggs used for analysis were laid between April 19 and May 14.

EGG RECORD OF HENS USED IN EXPERIMENT
Poor Producers—Low Rate of Laying

Hen number	Annual record	Eight weeks record
F 281	268	25
F 281	268	25
E 285	234	25
E 229	241	32
G 529	231	28
H 66	208	31
I 196	256	31
I 198	248	30
H 200	242	33
E 260	249	35

Good Producers—High Rate of Laying

Hen number	Annual record	Eight weeks record
F 554	234	47
H 581	213	44
I 790	204	44
H 253	246	45
H 91	182	41
H 235	......	45
G 512	238	38
I 402	......	42

It will be noted that in two places the annual egg record is not given. These hens were not trap-nested in the first year, but were from high-producing strain and made a good second-year record. It will also be noted that the low-rate layers were not especially poor, but made a normal production in the eight weeks. The good producers made more than normal production.

While a sufficient number of analyses have not yet been made, the results seem to show a possible relation between rate of laying and quality of the eggs, which in this case is the fat content. The following shows the percentage of ether extract (Fat) to entire egg for the three breeds.

Breed	High rate or good producers	Low rate or poor producers
Barred Plymouth Rocks	9.9015	13.3111
Oregons	10.4624	11.8718
Leghorns	10.6819	10.6749

Percent of ether extract to yolk:

	%	%
Barred Plymouth Rocks	29.4832	33.7508
Oregons	28.5055	31.0077
Leghorns	27.9141	28.4035

The analyses show that the Plymouth Rock eggs had the highest percentage of fat, the Oregons second, and the Leghorns third. The difference is small and probably without significance. It may be, however, that the hen with the largest supply of body fat is best able to maintain the fat content of the egg through a prolonged and heavy laying period, assuming that the fat of the eggs comes in a measure from the reserve fat of the body of the hen.

The results, however, seem to indicate some relation between rate of laying and the fat content of the egg. The amount of fat in the egg appears to be a variable factor, depending to some extent upon the rate of laying. It is another problem whether the food has any direct relation to the fat content of the egg. Final conclusions, however, cannot be drawn until further investigations have been made.

There was no question involved in these experiments as to good and poor layers. The hens used were all good layers, or had made high annual records in the previous year or years, but during the experiment some of them were laying slowly, or at a low rate, while others were laying heavily, and they were divided into two groups on this basis.

BIMONTHLY PRODUCTION IN RELATION TO ANNUAL PRODUCTION

The best layers as shown by their annual records were the best layers throughout the year. There are individual exceptions, but by grouping all the hens together, according to annual production, and averaging the groups, the results are very conclusive. The notion that poor layers lay well or as well as the good layers in the spring season, or season of maximum production, is clearly disproved. Not only are the poor layers poor at the beginning and end of the year, but they are poor at the middle, or the favorable season of production.

For the purpose of this study all the hens have been put into five groups: The first group laying 251 to 300 plus in the first laying year, the second 201 to 250 inclusive, the third 151 to 200, the fourth 101 to 150, the fifth 1 to 100 inclusive. The production has been summarized in bimonthly periods, beginning November and ending October. The total production within these periods or months does not always equal the total annual production, because the annual-production record dates from the date of the first egg, and the first egg may be laid at different times, sometimes in October or September, while others may not start until December or January, and the end of the laying year varies accordingly. The total production for the first year, as shown in the table, is the production for the twelve months of laying, beginning with the first egg. For this tabulation we have used the calendar year to show the distribution of production by bimonthly periods.

TABLE LV. BIMONTHLY PRODUCTION PER HEN. BARRED PLYMOUTH ROCKS

Group	No. hens	Average per hen						
		Nov. Dec.	Jan. Feb.	March April	May June	July Aug.	Sept. Oct.	First year
251-300	12	22.66	41.25	53.25	48.33	44.75	43.75	269.17
201-250	113	16.62	35.99	46.91	41.71	37.05	32.49	219.15
151-200	193	7.83	28.62	42.64	37.41	31.43	22.51	176.37
101-150	109	3.34	17.64	34.39	30.41	23.67	13.06	128.28
1-100	90	1.04	6.19	20.10	15.17	14.04	4.12	66.27
Mean	517	7.97	24.31	38.16	33.25	28.31	19.99	158.57

TABLE LVI. BIMONTHLY PRODUCTION PER HEN. WHITE LEGHORNS

Group	No. hens	Average per hen						
		Nov. Dec.	Jan. Feb.	March April	May June	July Aug.	Sept. Oct.	First year
251-300	33	32.97	41.45	51.94	47.09	46.42	38.21	266.78
201-250	137	26.82	35.87	47.02	43.22	38.74	27.92	224.63
151-200	112	13.88	25.64	42.75	39.02	29.93	19.64	176.77
101-150	64	6.42	18.61	35.67	31.72	21.28	9.36	128.06
1-100	40	1.65	7.10	20.15	16.95	12.60	4.32	64.20
Mean	386	17.60	27.54	41.54	37.70	31.24	20.88	181.71

TABLE LVII. BIMONTHLY PRODUCTION PER HEN. OREGONS

Group	No. hens	Average per hen						
		Nov. Dec.	Jan. Feb.	March April	May June	July Aug.	Sept. Oct.	First year
251-300	64	28.28	41.85	52.35	50.56	45.89	41.43	269.67
201-250	118	18.37	34.06	46.77	44.46	41.27	32.85	224.53
151-200	99	13.81	27.34	40.05	36.25	30.74	22.29	176.47
101-150	52	6.48	20.80	27.96	27.26	22.11	14.42	126.52
1-100	16	3.56	16.12	20.06	12.62	12.00	6.12	72.06
Mean	349	16.45	30.79	41.86	39.23	34.93	27.46	197.58

The average bimonthly production of the Barred Plymouth Rocks is given by groups in Table LV (see also Fig. 21). Among the 517 hens involved there were 12 with records above 250 eggs in the year; 113 averaged 201 to 250; 193 averaged 151 to 200; 109 averaged 101 to 150; and 90 averaged 1 to 100. The actual average production in the laying year is shown in the last column of the table. The annual production, however, as given is not the total of the bimonthly production in the table because the annual production of all our hens is figured, as already explained, from the date of the first egg to the end of the first twelve months.

The significant point in this summary is that the highest group has the highest average in any two months in the year beginning with November and December, and that there is a gradual decrease in bimonthly production as the annual production decreases.

There is a greater difference between good and poor layers in November and December than in any other two months figured on a percentage basis. For example, the average November and December production of highest group was 22.66 eggs a hen, while the production of the lowest group was 1.04 eggs, an actual difference of 21.62 eggs a hen. In March and April when all the groups lay the best there is an actual difference in the yield between the highest and lowest group of 33.15 eggs. Figured on a percentage basis, however, the increase from the lowest to the highest or the decrease from the highest to the lowest is greater in November and December than in any other two months.

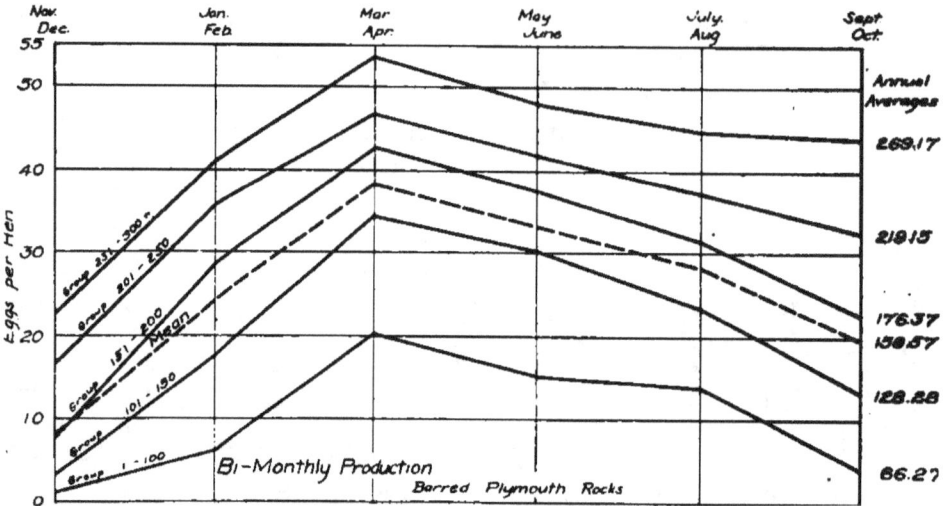

Fig. 21. Average bimonthly production of Plymouth Rock pullets, showing correlation between short-period production and annual production.

The relative values of good and poor layers are probably best shown by proportion or in ratio of production of birds in high groups to other groups. From Table LVIII it is seen that one hen in the highest group laid as many eggs in November and December as 21.79 hens in the lowest group, or 2.89 hens in the middle group. In March and April one hen in the highest group laid as many as 2.65 hens in the lowest group and 1.25 hens in middle group. In the last two months, September and October, the highest group averaged as many eggs per hen as 10.62 hens in the lowest group. Taking the annual average it required 4.06 hens or 4 times as many in the lowest group to lay as many eggs as 1 hen in the highest group. In the second lowest group, laying 100 to 150 eggs, it required 2.1 hens or double the number, to produce as many eggs in a year as the average hen in the first group averaging 250 to 300 eggs.

A summary of the bimonthly production of the White Leghorns is given in Table LVI (see also Fig. 22). In the highest group there were 33 hens that averaged 32.97 eggs a hen in November and December and

51.94 eggs in March and April. In the lowest group, laying fewer than 100 eggs, the production was only 1.65 eggs a hen in November and December and 20.15 eggs in March and April. At the end of the year or in the months of September and October the highest group averaged 38.21 eggs and the lowest 4.32 eggs.

TABLE LVIII. SHOWING PROPORTION BETWEEN PRODUCTION OF HENS IN HIGHEST GROUP AND THOSE IN EACH OTHER GROUP. BARRED PLYMOUTH ROCKS

Group	No. hens	Nov. Dec.	Jan. Feb.	March April	May June	July Aug.	Sept. Oct.	First year
251-300	12	1	1	1	1	1	1	1
201-250	113	1.36	1.15	1.14	1.16	1.20	1.35	1.23
151-200	193	2.89	1.44	1.25	1.29	1.42	1.95	1.53
101-150	109	6.78	2.34	1.55	1.59	1.89	3.35	2.10
1-100	90	21.79	6.66	2.65	3.19	3.19	10.62	4.06

There is a gradual and consistent decline in the number of eggs laid in different periods. This decline follows closely the decline in annual production. The fact stands out here that the best layers in the year are the best at any season of the year, even in March and April, the season of maximum production. While the difference in the egg yield in November and December and that of March and April between the highest group and the lowest is practically the same, the percentage difference is less in March and April than in November and December. In either period the low group could be culled out either in November and December or in March and April.

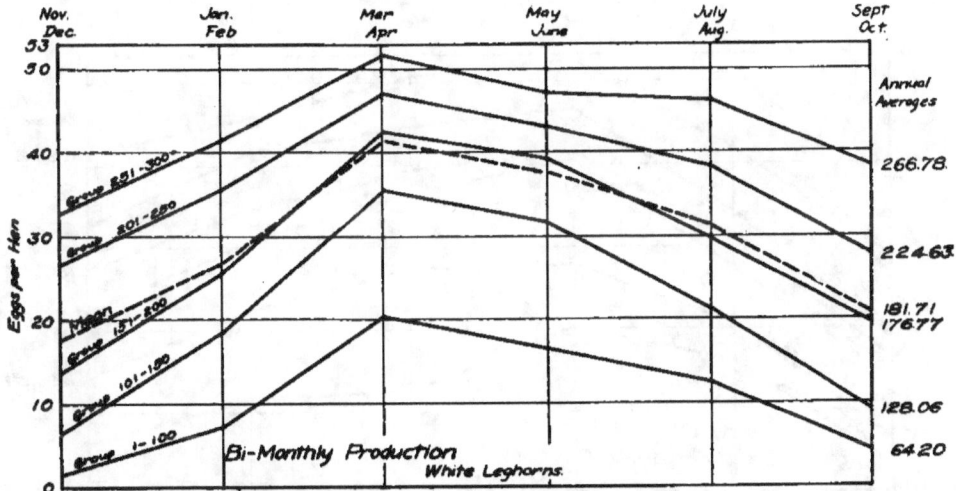

Fig. 22. Average bimonthly production of White Leghorn pullets, showing correlation between short-period production and annual production.

The relative value of hens of the different groups at different periods is better shown in Table LIX. For example, among the group of hens that laid 251 to 300 eggs in the year, one hen laid as many eggs in November and December as 19.98 hens in the lowest group averaged in the same month. In other words, 33 hens in the highest group laid 1088 eggs in November and December, while 40 hens in the low group laid only 66 eggs in the same time. So in other groups. Hen values

TABLE LIX. SHOWING PROPORTION BETWEEN PRODUCTION OF
HENS IN HIGHEST GROUP AND THOSE IN EACH
OTHER GROUP. WHITE LEGHORNS

Group	No. hens	Nov. Dec.	Jan. Feb.	March April	May June	July Aug.	Sept. Oct.	First year
251-300+	33	1	1	1	1	1	1	1
201-250	137	1.23	1.15	1.10	1.09	1.20	1.37	1.19
151-200	112	2.38	1.62	1.21	1.21	1.55	1.95	1.51
101-150	64	5.13	2.23	1.46	1.48	2.18	4.08	2.08
1-100	40	19.98	5.84	2.58	2.78	3.68	8.84	4.16

decrease in bimonthly periods especially in November and December as their annual production decreases.

The same thing is true of the Leghorns as of the other breeds. In this regard, there is less difference in hen values in the months of March and April, the season of highest production, than in any other period. In these two months one hen in the highest group equaled 2.58 hens in the lowest group.

In the total for the year for Leghorns, one hen in the highest group equaled 4.16 hens in the lowest group. In other words, each of four and a fraction hens in the lowest class took up as much room and ate about as much feed in the year as one hen did in the high group. Knowing the production of the hens in March and April an intelligent culling may be done.

In the case of the Oregons (Table LVII; see also Fig. 23), there were 64 hens in the highest group; 118 in the second; 99 in the third; 52 in the fourth; and 16 in the fifth or a total of 349 hens. Sixty-four

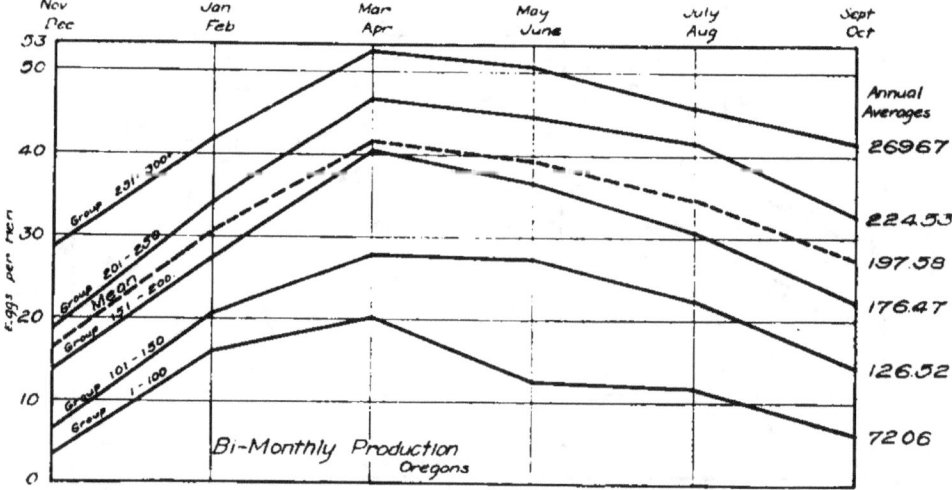

Fig. 23. Average bimonthly production of Oregons, showing correlation between short-period production and annual production.

hens in the highest group averaged 28 eggs in the bimonthly period of November and December, and 52.35 eggs in March and April. In the lowest group sixteen hens averaged 3.56 eggs in November and December and 20.06 eggs in March and April, and so on.

The highest group laid in November and December 28.28; 24.72 eggs more than the lowest group, or some seven hundred percent more. In March and April the highest group laid 32.29 eggs more than the lowest, or some 160 percent more. The actual difference in eggs laid was greater in the high months than the low months, but the percentage difference was greater in the low months.

Computing the difference in another way, as in Table LX, one hen in the highest group laid as many eggs in November and December as did 7.94 hens in the lowest group, while in March and April 2.61 hens in the lowest group laid as many as one hen in the highest. The same thing is true in all of the groups and all the months.

TABLE LX. SHOWING PROPORTION BETWEEN PRODUCTION OF HENS IN HIGHEST GROUP AND THOSE IN EACH OTHER GROUP. OREGONS

Group	No. hens	Nov. Dec.	Jan. Feb.	March April	May June	July Aug.	Sept. Oct.	First year
251-300	64	1	1	1	1	1	1	1
201-250	118	1.54	1.23	1.12	1.14	1.11	1.26	1.20
151-200	99	2.04	1.53	1.31	1.39	1.49	1.86	1.53
101-150	52	4.36	2.01	1.87	1.85	2.08	2.87	2.13
1-100	16	7.94	2.59	2.61	4.01	3.82	6.77	3.74

In the discussion of short-period production as a measure of laying capacity, it was shown that as the maximum best-two-months production increased, the annual production increased. In this study we see that as the annual production increased, the best two-months production increased. While the greatest increase in production has come in the low months of the year, or at the beginning and end of the year, there is also a considerable increase in production in the favorable months.

Though the production draws closer together in the spring months, than at the beginning and end of the year, there is in these two months a decided difference between good and poor layers. In other words, there is a strong correlation between laying capacity, as determined by the annual record, and March and April production.

The difference in production in November and December, however, is not always a question of rate of laying. Where the records show that a hen laid only 3.66 eggs as an average in November and December, it does not mean that the production was spread over the two-months period. It means that, while the rate of laying was low, the pullets did not start as early as the high group. But in March and April when the hens were all laying, and laying throughout the full two-months period, the difference in production in the group is a difference in rate of laying. In the first two months, there is a question of laying maturity involved as well as rate of laying, while in March and April, the only question is rate of laying.

Bimonthly Percentage Production. In Tables LXI, LXII, LXIII is shown in percentage the bimonthly production of Barred Plymouth Rocks, White Leghorns, and Oregons, arranged in groups according to annual production. This is based on the possible hen production of an egg a day. It will be understood that these bimonthly percentages would vary if artificial lights were used on the flocks. The production would be higher in the fall and early winter months with lights and probably less in the spring months. But under natural conditions of lighting and care, the table affords a basis for estimating fairly accurately the annual production of a flock. In the absence of trap-nest records, the owner of a flock under good care may know, within a reasonable range, from the flock record in any two months, preferably in the spring months of best production, what the production of the flock will be in the year. If he has a good laying strain and they do not come up to the production shown in the table in any months he will know that he is not giving them proper care.

TABLE LXI. PERCENTAGE BIMONTHLY PRODUCTION. BARRED PLYMOUTH ROCKS

Group	No. Hens	Nov. Dec.	Jan. Feb.	March April	May June	July Aug.	Sept. Oct.	First year	Annual Prod.
251-300	12	37.16	69.92	87.29	79.23	72.17	71.72	73.74	269.17
201-250	113	27.24	61.00	76.90	68.37	59.76	53.27	60.04	219.15
151-200	193	12.83	48.51	69.88	61.32	50.69	36.89	48.32	176.37
101-150	109	5.47	29.90	56.38	49.85	38.18	21.40	35.14	128.28
1-100	90	1.71	10.49	32.95	24.86	22.65	6.75	18.16	66.27

TABLE LXII. PERCENTAGE BIMONTHLY PRODUCTION. WHITE LEGHORNS

Group	No. hens	Nov. Dec.	Jan. Feb.	March April	May June	July Aug.	Sept. Oct.	First year	Annual Prod..
251-300	33	54.05	70.26	85.19	77.19	74.88	62.64	73.09	266.78
201-250	137	43.96	60.79	77.09	70.85	62.48	45.78	61.54	224.63
151-200	112	22.76	43.46	70.08	63.96	48.27	32.20	48.43	176.77
101-150	64	10.52	31.54	58.48	51.97	34.32	15.34	35.09	128.06
1-100	40	2.71	12.03	33.03	27.79	20.32	7.09	17.59	64.20

TABLE LXIII. PERCENTAGE BIMONTHLY PRODUCTION. OREGONS

Group	No. hens	Nov. Dec.	Jan. Feb.	March April	May June	July Aug.	Sept. Oct.	First year	Annual Prod.
251 300	64	46.36	70.95	85.83	82.89	74.02	67.93	73.88	269.67
201-250	118	30.12	57.73	76.67	72.89	66.57	53.86	61.52	224.53
151-200	99	22.65	46.34	65.66	59.43	49.58	36.55	48.35	176.47
101-150	52	10.62	35.27	45.84	44.70	35.67	23.64	34.66	126.52
1-100	16	5.84	27.33	32.89	20.69	19.35	10.04	19.74	72.06

If, for example, his flock is laying about 86 percent in March and April it has a capacity of more than 250 eggs a hen, and if given good care he may expect a production in the year of more than 250 eggs, a hen, though there may be no flocks of large size that would average that number of eggs.

In the experimental flocks the March and April production of hens laying above 250 eggs, was as follows:

	March and April %	Annual eggs
Barred Plymouth Rocks	87.29	269.17
White Leghorns	85.19	266.78
Oregons	85.83	269.67

If the production is 32 to 33 percent in March and April he should expect a production of about 70 eggs a hen in the year. If the production in July and August is about 50 percent he can be assured of a production of about 176 eggs.

The greatest uniformity in the three breeds is in March and April, indicating that this is the best period for estimating the annual production of the flock. A 60-percent production in March and April indicates a satisfactory egg yield, while a production above 70 percent indicates a production above 200 eggs a hen.

Percentage Deviation from the Mean. In Tables LXIV, LXV, and LXVI, is given by groups the percentage deviation from the mean bimonthly production. The mean production is shown in eggs per hen, and the deviation in percentage in bimonthly periods is shown for the

TABLE LXIV. PERCENTAGE DEVIATION FROM THE MEAN. BARRED PLYMOUTH ROCKS

| Group | No. hens | Bimonthly production | | | | | | |
		Nov. Dec.	Jan. Feb.	Mar. Apr.	May June	July Aug.	Sept. Oct.	First year
Mean........	......	7.97	24.31	38.16	33.25	28.31	19.99	158.57
251–300....	12	+184.32	+69.68	+39.54	+45.35	+58.07	+118.86	+69.75
201–250....	113	+108.53	+48.05	+22.93	+25.44	+30.87	+62.53	+38.20
151–200....	193	−1.76	+17.73	+11.74	+12.51	+11.02	+12.60	+11.23
101–150....	109	−58.09	−27.44	−9.88	−8.54	−16.39	−34.67	−19.10
1–100....	90	−86.95	−74.54	−47.33	−54.38	−50.41	−79.39	−58.21

TABLE LXV. PERCENTAGE DEVIATION FROM MEAN. WHITE LEGHORNS

| Group | No. hens | Bimonthly production | | | | | | |
		Nov. Dec.	Jan. Feb.	Mar. Apr.	May June	July Aug.	Sept. Oct.	First year
Mean........	......	17.60	27.54	41.54	37.70	31.24	20.88	181.71
251–300+	33	+87.33	+50.51	+25.05	+24.91	+48.59	+83.00	+46.82
201–250....	137	+52.39	+30.25	+13.20	+14.64	+24.01	+33.72	+23.62
151–200....	112	−21.14	−6.90	+2.91	+3.50	−4.19	−5.94	−2.72
101–150....	64	−63.52	−32.42	−14.14	−15.86	−31.88	−55.17	−29.53
1–100....	40	−90.63	−74.22	−51.53	−55.04	−59.67	−79.31	−64.67

TABLE LXVI. PERCENTAGE DEVIATION FROM MEAN. OREGONS

| Group | No. hens | Bimonthly production | | | | | | |
		Nov. Dec.	Jan. Feb.	Mar. Apr.	May June	July Aug.	Sept. Oct.	First year
Mean........	......	16.45	30.79	41.86	39.23	34.93	27.46	197.58
251–300....	64	+71.92	+35.92	+25.06	+28.88	+31.38	+50.88	+36.50
201–250....	118	+11.68	+10.62	+11.73	+13.33	+18.15	+19.63	+13.66
151–200....	99	−16.05	−11.21	−4.33	−7.59	−11.99	−18.83	−10.67
101–150....	52	−60.67	−32.45	−33.21	−30.51	−36.70	−47.48	−35.96
1–100....	16	−78.36	−47.65	−69.85	−67.83	−65.64	−77.71	−63.52

different groups. In the case of the Barred Plymouth Rocks, for example, the mean for November and December is 7.97 eggs per hen. In the highest group the production is 184.32 percent above the mean. In the second group it is 108.53 percent above the mean; in the third group it is 1.76 below the mean, while in the lowest group it is 86.95 percent below the mean. The plus sign before the number indicates above the mean; the minus sign indicates below the mean.

In March and April the mean is 38.16 eggs for Barred Rocks, while in the highest group the production is 39.54 percent above the mean. In the lowest group it is 47.33 percent below the mean. In the case of the Oregons, the mean was 16.45 eggs a hen in November and December, while the highest group laid 71.92 percent more than the mean, and the lowest group 78.36 less than the mean. In March and April the mean is 41.86 eggs a hen, while the highest group laid 25.06 percent more than the mean and the lowest 69.85 less than the mean.

The significance of this tabulation is that while the difference in the number of eggs laid is greater in March and April than in November and December, between the highest and lowest group the percentage difference is less in March and April than in November and December. There is not as wide a gap between the high and low groups in March and April when the difference is figured in percentage above and below the mean.

Conclusions. To summarize some of the results: When a flock averages high in November and December, or say 20 eggs a hen, which is the average of the three breeds for the group laying 201 to 250, we can expect a yield of more than 200 eggs in a year. Judging from the average of the highest group in the three breeds, a production of 27.75 eggs in November and December would indicate an annual production of 251 to 300 plus. When they lay fewer than 4 eggs in November and December, we may expect a yield of fewer than 100 a year. The intermediate groups show corresponding results.

The hens showing the lowest production in November and December, not only show the lowest through all the months, but when the year's total is figured up, it required 3.74 hens in the lowest group to lay as many in the year as one hen in the high group. In March and April one hen in the high group was equal to 2.61 hens in the lowest group. This indicates that culling may be fairly accurately done in March and April as well as November and December.

WINTER PRODUCTION AND ANNUAL PRODUCTION

In the production of the first quarter of the laying year referred to as winter production, the results show very conclusively that low production in these months, means low production for the rest of the year. High production in the winter months means high production for the rest of the year. Selective culling may be profitably done on the basis of first three months production.

The average results for the different years are given below.

Breed	No. of hens	Winter production to Jan. 31	Rest of year	Annual production
Barred Plymouth Rocks	517	19.79	138.78	158.57
Leghorns	356	35.44	146.27	181.71
Oregons	349	37.02	160.57	197.58

(Unidentified or floor eggs are not included.)

When the hens are grouped according to annual production and breed, it is shown that the best layers in the first quarter are the best in the rest of the year. For example, in the table for Barred Plymouth Rocks, the hens laying between 250 and 300 averaged 44.25 eggs a hen to January 31, and 224.92 in the rest of the year. The poor group laying from 1 to 100 averaged 3.17 eggs to January 31, and 63.08 in the rest of the year.

The same results are secured with Leghorns and Oregons, as shown in Table LXVII.

The characteristics of the good layers as shown by this tabulation, are good fall and early winter production, with good production throughout the rest of the year, including good production in the late summer and early fall. In other words, by breeding the length of the laying period is extended, both at the beginning and end of the laying year.

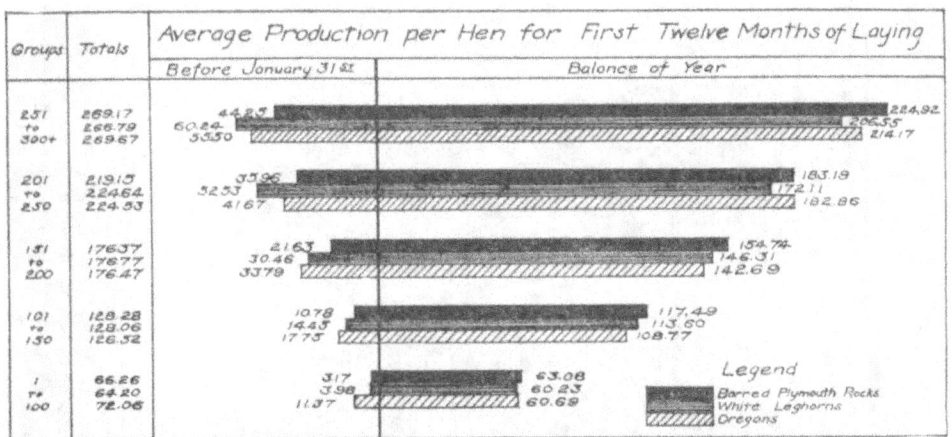

Fig. 24. Hens grouped according to production, showing correlation between fall and winter production and the remainder of the year.

CORRELATION BETWEEN THE PRODUCTION OF THE FIRST YEAR AND THAT OF THE SECOND AND SUBSEQUENT YEARS

The problems discussed so far in this bulletin have had reference only to the pullet year or first year's production. The conclusions drawn are based on those records. It is of practical importance to know whether high fecundity is inherited in the second and subsequent years. A profitable hen must be profitable for more than one year. If a hen produces well in her first year, will she produce well in subsequent years, or is it true that the more she lays in her first year, the fewer she will lay in subsequent years? Again, it is highly important for profitable egg production, that the production period of the hen, or the profitable laying life of the hen, be increased.

It was the practice each year in our experiments to retain the best layers in the first year for the second year, and the best of these again for the third and subsequent years. A few of the best layers have reached nine years of age with continuous trap-nest records. Some poor layers in the first year were also retained in second and subsequent years.

The trap-nest record begins when the hen first begins to lay, and the first year dates from that time. The second year begins at the end of the first twelve months of laying, and so on for subsequent years.

86

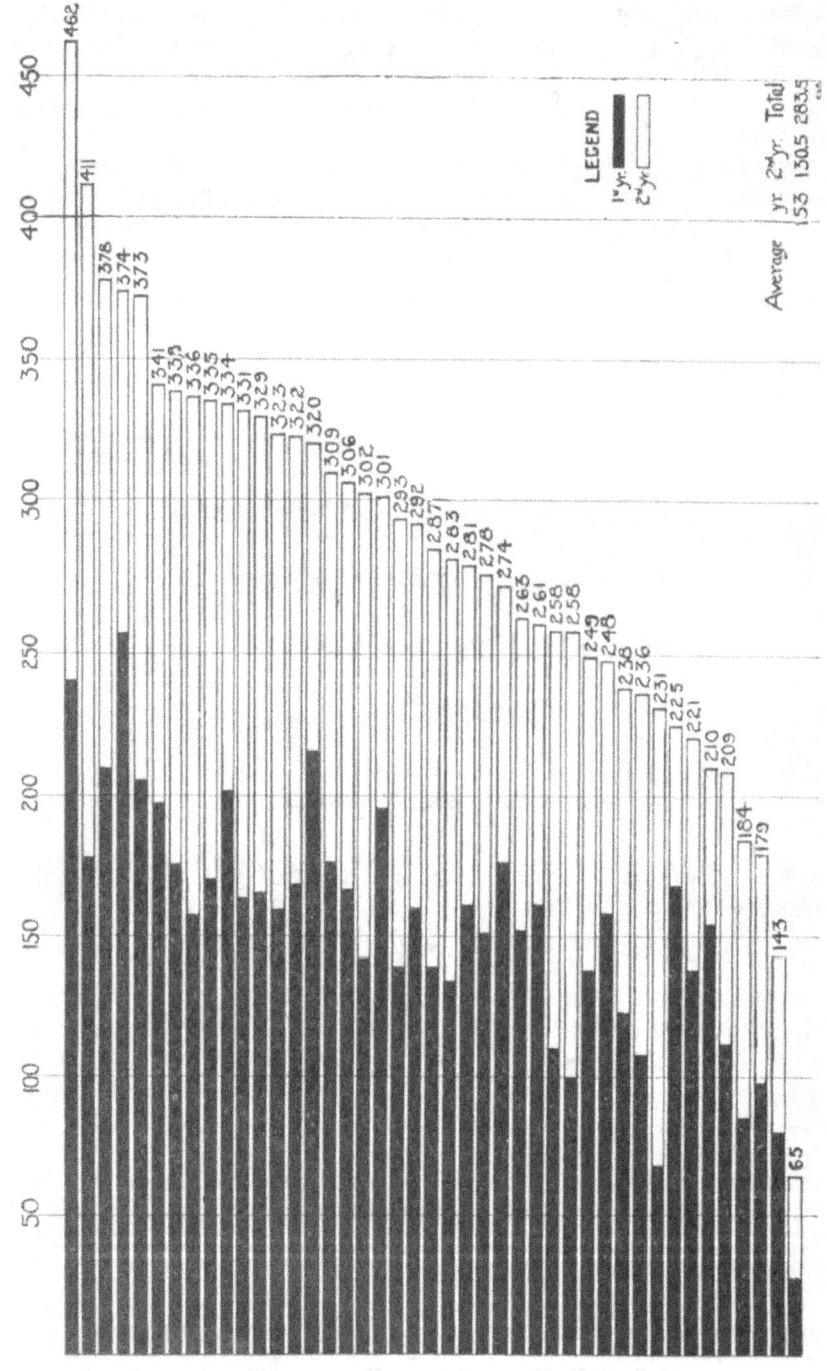

Fig. 25. Graph representing two years' trap-nest records of a pen of forty-three Leghorns and first crosses (White Leghorns and Barred Plymouth Rocks) grouped according to two years' egg production. The first and second year's production is given at the left of the chart and the total at the end of the lines on the right. This graph shows that the best layers in the first year, represented by the blank lines, are on the average the best in two years.

The results show clearly: (1) The first year is on the average the most productive or profitable year in the life of the hen. (2) The best layers in the first year are, on the average, the best in the second and subsequent years. In other words, there is a clear correlation between the production of the first year and the second and subsequent years. (3) The decrease in production from the first to the second and subsequent years is greater in the case of high producers than low first-year producers. But the second year's production of good producers is better than that of the poor producers. (4) The very poorest producers in the first year, on the average, lay better in the second year than the first year.

Averaging all the hens with two-year complete records for the different years the production is as follows:

Breed	No. of hens	1st Yr.	2d Yr.	Decrease	Percent decrease
Barred Plymouth Rocks	157	178.77	128.70	50.07	28.01
Leghorns	185	209.16	163.65	45.51	21.76
Oregons	219	200.49	161.95	38.54	19.22

This shows a considerable decrease from first to second year in all three breeds. It is understood that the hens represented in the above tabulation are of good laying strains, but there was a wide range in production of individual hens. As shown in other tables, this variation has extended from less than one hundred eggs to three hundred eggs a hen.

Computations have also been made for hens with three-year records. Their averages are as follows:

Breed	No. of hens	1st Yr.	2d Yr.	3d Yr.	Decrease 1st.-3d	Percent decrease
Barred Plymouth Rocks	35	225.89	158.20	124.20	101.69	45.02
Leghorns	96	227.58	174.59	157.24	70.34	30.91
Oregons	74	231.01	189.78	158.96	72.05	31.19

It is seen that the decrease continues from the second to the third year, though it is not quite as much in the third year as in the second. In the case of the Barred Plymouth Rocks, there is a decrease of over 100 eggs, or 45.02 percent, from first to the third year. The decrease for the Leghorns and Oregons is 70.34 and 72.05 eggs a hen. It will be remembered that the hens represented here are all heavy producers in the first year.

The Barred Plymouth Rocks decreased more rapidly in the second year and third year than the Leghorns and Oregons. The Leghorns and Oregons showed about the same decrease. Whether this is a breed or a strain difference is not clear, but it would appear from these experiments that the heavier breeds do not maintain profitable production for as many years as the lighter breeds.

To learn whether good layers and poor layers show the same percentage decrease each year, all the hens with two-year and three-year records have been grouped according to production in the first year.

The grouping of the Barred Plymouth Rocks (Table LXVIII) for the first and second years' production shows that the hens that laid 250 to 300-plus eggs in the year, averaging 266.43, averaged in the second year 164.28 eggs, a decrease of 102.15 eggs, or a percentage decrease of 38.34.

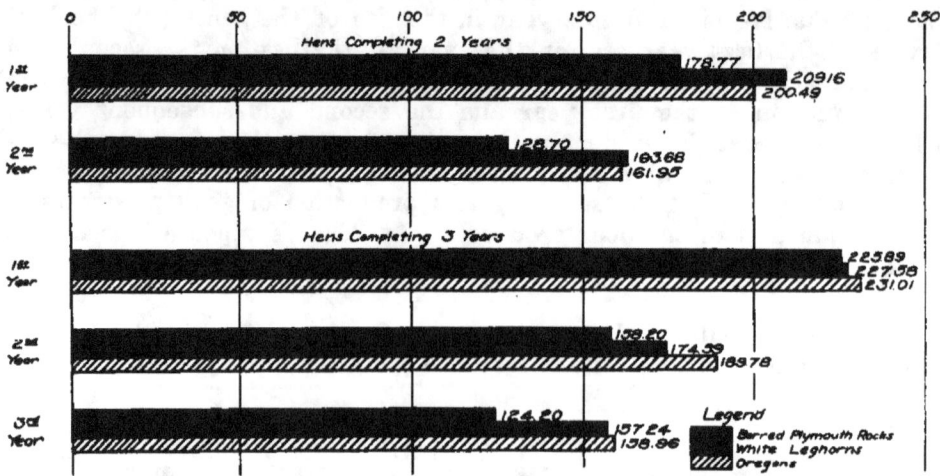

Fig. 26. Average production per hen of Barred Plymouth Rocks, White Leghorns, and Oregons, showing correlation between first and second years' production of the same hens, and first, second, and third years' production of the same hens.

While the decrease is greatest in the highest group the second year's production is highest in the highest group. In the lowest group laying one hundred eggs or less, and averaging 72.21 eggs, the average for the second year was 77.78, or an increase of 5.57 eggs a hen, or a percentage increase of 7.71.

TABLE LXVII. SUMMARY

Groups	No. of hens	Average production per hen*		
		Through January	Rest of year	Annual
Barred Plymouth Rocks				
251 to 300+	12	44.25	224.92	269.17
201 to 250	113	35.96	183.19	219.15
151 to 200	193	21.63	154.74	176.37
101 to 150	109	10.78	117.49	128.28
1 to 100	90	3.17	63.08	66.26
Average		19.79	138.78	158.57
White Leghorns				
251 to 300+	33	60.24	206.55	266.79
201 to 250	137	52.53	172.11	224.64
151 to 200	112	30.46	146.31	176.77
101 to 150	64	14.45	113.60	128.06
1 to 100	40	3.98	60.23	64.20
Average		35.44	146.27	181.71
Oregons				
251 to 300+	64	55.50	214.17	269.67
201 to 250	118	41.67	182.86	224.53
151 to 200	99	33.79	142.69	176.47
101 to 150	52	17.75	108.77	126.52
1 to 100	16	11.37	60.69	72.06
Average		37.02	160.57	197.58

*Unidentified or "floor" eggs not counted.

The same law of diminishing returns holds true with Leghorns and Oregons, that the higher the first year's production the greater the decrease in the second, but the second year's record of the highest group is better than the second year's record of the lower groups. If, therefore, high fecundity is inherited in the first year, it is inherited in the second. In Tables LXXI, LXXII, and LXXIII, showing first, second, and third years' records of hens completing three years, it will be noted that there is a decrease in all groups from the second year to the third, with one exception, that of the lowest group of Oregons. As it happened that there was only one hen in that " group," no significance can be attached to this case.

TABLE LXVIII. RATE OF DECREASE IN PRODUCTION—2 YEARS SUMMARY. BARRED PLYMOUTH ROCKS

Groups	No. hens	Average 1st year	Average 2d year	Decrease	Percent decrease
251 to 300+	7	266.43	164.28	102.15	38.34
201 to 250	69	221.36	144.07	77.29	34.92
151 to 200	36	180.11	134.05	46.06	25.58
101 to 150	22	126.54	109.09	17.45	13.79
1 to 100	23	72.21	77.78	−5.57	−7.71
Average	157	178.77	128.70	50.07	28.01

TABLE LXIX. RATE OF DECREASE IN PRODUCTION—2 YEARS SUMMARY. WHITE LEGHORNS

Groups	No. hens	Average 1st year	Average 2d year	Decrease	Percent decrease
251 to 300	27	266.81	169.00	97.81	36.66
201 to 250	96	227.57	169.54	58.03	25.94
151 to 200	39	180.25	165.46	14.79	8.20
101 to 150	19	127.31	136.84	−9.53	−7.49
1 to 100	4	94.00	97.25	−3.25	−3.46
Average	185	209.16	163.68	45.48	21.74

TABLE LXX. RATE OF DECREASE IN PRODUCTION—2 YEARS SUMMARY. OREGONS

Groups	No. hens	Average 1st year	Average 2d year	Decrease	Percent decrease
251 to 300+	51	268.27	188.45	79.82	29.76
201 to 250	73	223.93	179.01	44.92	20.06
151 to 200	49	173.89	146.59	27.30	15.69
101 to 150	37	125.02	126.46	−1.44	−1.15
1 to 100	9	81.33	102.88	−21.55	−27.12
Average	219	200.49	161.95	38.54	19.22

It will also be noted that the decrease among the different groups from the second year to the third year is not as consistent as from the first to the second. This would be expected from the relatively small number of hens with complete three-year record. From the average of the three breeds, however, the decrease is greatest from second to third year among the higher group of layers, but the decrease in the third year is much less than in the second year.

The grouping given in these tables includes hens of different yards in the different years with records for two and three years. The record shows the same general decrease in different yards of different years. The production of the first year's stock, 1908-09, does not show much decrease in the second year, but this agrees with the result of all the years, that the low-first-year hens do not show as much decrease as the

TABLE LXXI. SUMMARY OF RATE OF DECREASE IN PRODUCTION. BARRED PLYMOUTH ROCKS

Group	No. hens	Average 1st year	Average 2d year	Decrease from 1st year	Percent decrease from 1st year	Average 3d year	Decrease from 1st year	Percent decrease from 1st year	Decrease from 2d year	Percent decrease from 2d year
251 to 300+	6	269.00	182.67	86.33	32.09	157.00	112.00	41.64	25.67	14.05
201 to 250	25	223.20	153.20	70.00	31.36	119.92	103.28	46.27	33.28	21.72
151 to 200	4	178.00	151.50	26.50	14.89	141.50	36.50	20.51	10.00	6.61
Average	35	225.89	158.20	67.69	29.97	124.20	101.69	45.02	34.00	21.49

TABLE LXXII. SUMMARY OF RATE OF DECREASE IN PRODUCTION. WHITE LEGHORNS

Group	No. hens	Average 1st year	Average 2d year	Decrease from 1st year	Percent decrease from 1st year	Average 3d year	Decrease from 1st year	Percent decrease from 1st year	Decrease from 2d year	Percent decrease from 2d year
251 to 300+	19	266.32	171.79	94.53	35.50	148.74	117.58	44.15	54.21	31.56
201 to 250	60	229.27	175.12	54.15	23.62	161.33	67.94	29.63	13.79	7.87
151 to 200	15	185.80	181.40	4.40	00.24	155.60	30.20	1.63	25.80	14.19
101 to 150	2	122.50	134.50	-12.00	-8.92	127.50	-5.00	-3.92	7.00	5.20
Average	96	227.58	174.59	52.99	23.28	157.24	70.34	30.91	17.35	9.94

TABLE LXXIII. SUMMARY OF RATE OF DECREASE IN PRODUCTION. OREGONS

Group	No. hens	Average 1st year	Average 2d year	Decrease from 1st year	Percent decrease from 1st year	Average 3d year	Decrease from 1st year	Percent decrease from 1st year	Decrease from 2d year	Percent decrease from 2d year
251 to 300+	31	270.68	196.48	74.20	27.41	170.77	99.91	36.91	25.71	13.08
201 to 250	28	231.36	198.32	33.04	14.28	153.57	77.79	32.76	44.75	22.56
151 to 200	8	175.38	184.75	-9.37	-5.07	165.13	10.25	5.84	19.62	10.62
101 to 150	6	132.50	147.50	-15.00	-11.32	132.16	00.34	00.26	15.34	10.40
1 to 100	1	28.00	37.00	-9.00	-24.32	55.00	-27.00	-49.09	-18.00	-32.73
Average	74	231.01	189.78	41.23	17.85	158.96	72.05	31.19	30.62	16.13

high hens, and the very lowest show an increase in the second year over the first. Out of 92 Barred Plymouth Rocks, in the first year averaging 86.14 eggs, 59 of the best were kept over into the second year. The average of the 59 in their first year was 89.50; in the second year the average was 78.61.

Of the 50 Leghorns averaging 106.88 in the first year, 28 of the best averaging 120.89 in the first year averaged in the second year 97.46.

This low second-year production of the two breeds in connection with their first-year production is additional proof that the foundation stock in these experiments was of low fecundity.

Of 561 hens with complete records for two years, there were 104 that laid more eggs the second year than the first. In the first year, these averaged 135.2 eggs, and in the second year 157.25 eggs. In the case of 35 of these hens, the records show that they started laying early and moulted in the fall and winter. This would account for a lower first-year record than the second. Of 25 others, the record shows that they started to lay late, most of them being late hatched. The fact that the 104 hens laid more in the second year than the first, therefore, may be explained in part at least by the fact that in a large number of cases environmental conditions were not favorable for high first-year production.

Limit of Production or Laying Longevity. The statement has been made by numerous authors that 600 eggs was the limit of production of a hen in a lifetime. This theory seems to have originated with a writer named Geyelin who said, "It has been ascertained that the ovarium of a fowl is composed of 600 ovula or eggs. Therefore, a hen during the whole of her lifetime cannot possibly lay more eggs than 600." This statement has been disproved by trap-nest records at this Station. Many of the station hens have laid more than 600 eggs, and a number have laid more than double 600.

Table No. LXXIV gives a list of twenty hens bred at this Station that have laid, up to the end of 1920, more than 1000 eggs. This table gives the hen number, her date hatched, and the egg record for each year. It also gives her pedigree or ancestry so far as known. It will be noted that practically all of them have high-record ancestors. It will also be noted that these hens are either direct descendants of or were closely related to 1000-egg hens. Hen C521 (Lady MacDuff) is the dam of four of these 1000-egg hens and the sire's dam of another. Lady MacDuff herself did not quite reach 1000 eggs before death, but a full sister (C547) has a record up to the end of her eighth year of 1300 eggs, which is the highest record so far secured. This hen in her ninth year is still laying well.* White Leghorn hen A27, record 1188 eggs in eight years, is the dam of two other 1000-egg hens and the sire's dam of five other 1000-egg hens. In fact, all the Leghorn 1000-egg hens are related to A27, and some of the Oregon hens of 1000-egg records carry some of her blood in the early generations.

A study of this table would indicate that the characteristic of long laying is inherited. The practical significance of the record will be apparent from the average production of the twenty hens. In the first year, the average was 235.75. There is a gradual decrease each year for five years, the average for the fifth year being 163.2. That means five

*Since this text was written C547 died (June 11, 1921). The complete record for her entire life was 1335 eggs.

TABLE LXXIV. HENS LAYING ONE THOUSAND EGGS OR MORE

Hen No.	Breed	Date hatched	1	2	3	4	5	6	7	8	9	10	Total	Dam	Dam's dam	Dam's dam's dam	Dam's sire's dam	Dam's sire's dam	Sire's dam	Sire's dam's dam	Sire's sire's dam	Sire's sire's dam	Sire's dam's sire's dam	Sire's dam's dam
A27 ..	W. L.	2- 4-10	240	222	202	155	168	139	61	1+	...	...	1188	...	...	...	...	...	...	...	...	...	...	...
A60 ..	Ore.	2-19-10	177	234	226	179	142	115	62	24+	...	...	1159	A 27 *240	...	...	...	...	...	...	...	...	...	...
H83N	W. L.	8-21-11	139	197	200	181	179	119	81+	...	...	...	1096	...	250 200 202	...	...	...	...	...	...	...	...	...
B14 .	Ore.	3-19-11	215	206	208	198	189	148	81	31	2+	...	1278	Yd 8 201	250 200 202	...	...	...	...	...	...	...	...	...
B42 ..	Ore.	2- 5-11	228	250	184	171	135	105	119	35	...	...	1282	A 66 201	O 34 229	...	...	...	O 31 229	...	...	...	...	...
C547	Ore.	4-29-12	250	225	211	192	156	108	76	79	54	1+	1300	B 13 206 226	B9 180 203	...	...	...	A 27 240	...	...	...	...	...
D328 .	Ore.	3-20-13	215	211	210	183	145	94+	...	...	...	...	1061	B8 246	...	...	...	...	A 27 240	...	...	...	...	...
D351 .	Ore.	3- 7-13	247	210	192	149	141	84	...	...	...	...	1023	B8 246	...	...	...	...	A 27 240	...	...	...	...	...
E21 ..	W. L.	3-15-14	259	249	172	215	193	127+	...	...	...	...	1215	B 12 251	O 34 229	O 34 229	A 27 240	O 34 229	A 27 240	...	...	...	...	...
E37 ..	W. L.	5-18-14	217	232	134	129	200	112	...	...	...	...	1024	D399 141+ D 203	B9 180 203	...	A 27 240	O 34 229	...	...	...	...	...	...
E40 ..	W. L.	5-27-14	245	218	172	191	153	61+	...	...	...	...	1040	B 12 251	O 34 229	...	...	...	A 27 240 C543 291	A 27 240	O 34 229	A 27 240	...	...
E115 .	Ore.	4- 3-14	301	190	205	172	151	1+	...	...	...	...	1020	B 42 228 250	A 66 201	...	O 34 229	...	...	...	A 27 240	O 34 229	...	250 200 202
E215 ..	Ore.	5-11-14	283	194	191	187	172	132	...	...	...	...	1159	C521 303	A 66 201	...	O 34 229	...	B 42 228 250	...	A 27 240	O 34 229	...	...
E218 .	Ore.	5-21-14	258	214	177	179	187+	...	...	...	...	...	1015	C521 303	A 66 201	...	O 34 229	...	B 42 228 250	...	A 27 240	O 34 229	...	...
E260 .	W. L.	5-27-14	249	183	195	149	165	112	...	...	...	...	1053	A 27 240	...	...	...	...	O 34 229	...	A 27 240	O 34 229	...	...

E285	Ore.	3- 5-14	234	173	191	181	122	130	.....	.....	.....	1031	C521 303	A 66 201	O34 229	.....	B 42 228 250	A27 240	O34 229	.....		
E291	Ore.	3-16-14	224	228	181	151	135	97	.....	.....	.....	1016	C521 303	A 6ᶜ 201	O34 229	.....	B 42 228 250	A27 240	O34 229	.....		
E315	W. L.	7-14-14	228	186	178	148	147	135	.....	.....	.....	1022	C516 267	A 45 215	O34 229	.....	A 27 240	.....	.....	.....		
F284	Ore.	1- 6-15	235	210	227	192	213	.....	.....	.....	.....	1077	Yd. M Yd.	.....	.....	C521 303	A 66 201	B42 228 250	A27 240	O34 229		
F621	W. L.	3- 2-15	271	223	203	177	171	.....	.....	.....	.....	1045	O	.....	.....	D718 254	C515 241	.....	A27 240	A27 240		
Average			235.75	212.9	193.1	173.95	163.2	107	80	34	28	.....										

* In the pedigree the figures under the hen numbers represent the first-year egg production. In some cases the second year is also given.

years of profitable production. The record of the sixth year, 107 eggs, is considerably above the average production of the hens of the United States. If the flocks of the country can be bred up so that their profitable period of production will be extended from about two years to four or five years, it would mean a great saving in the cost of production. The cost of reproducing the flock, including incubation and rearing, every two years is very great. By breeding from the best layers, the poultry breeder is taking a certain method of reducing the cost of production by lengthening the profitable life of the hen.

SUMMARY

1. Individual variation in egg production is very great, ranging in our experiments from 0 to more than 300 eggs in a year.

2. This variation does not appear to be a breed characteristic.

3. The highest individual record among the Barred Plymouth Rocks was 308 and the lowest 3 eggs. The highest White Leghorn record was 302 and the lowest 1 egg. The highest among the Oregons was 309 and the lowest 14 eggs.

4. There was apparently no decrease in variation when the flock production was increased by breeding. The range between the high and the low remains about the same.

5. The highest egg record in the original or foundation stock was 218 eggs. This has been increased to 308 as the maximum in the pedigreed, high-producing stock.

6. In a supplementary experiment in a flock of the Station's strain of Oregons, kept at the Oregon State Hospital, Salem, a high record of 330 eggs was secured with a total of 17 hens laying 300 eggs or more.

7. The pullet progeny of high-record, pedigreed stock, showed a large increase in production over the unselected, low-producing foundation stock.

8. Regardless of any question of prepotency, the selection of breeding stock on the basis of high production record is a certain method of increasing production.

9. Some individuals, however, showed greater power of transmitting high fecundity than others of the same breeding.

10. Good layers were not always produced by good layers, nor were poor layers always produced by poor layers.

11. More rapid progress will be made by the breeder if he can test the breeding quality of his stock, and use for breeding those hens and males whose progeny has shown high production.

12. In a few exceptional cases, the production of the pullet progeny was higher than that of the dam and sire's dam.

13. The average production of the pullets was less than the average of the parents where the parents were selected among the highest producers, but there were individual cases where the pullet records were higher than the parents' records.

14. Breeding from the highest producers decreased the number of poor producers, but it did not decrease variability nor obviate the necessity for continuing selection of breeding stock.

15. Breeding from the highest producers showed a progressive increase in the maximum individual production, showing that the opportunity for selection was as great at the end as at the beginning of the experiment.

16. Under favorable environmental conditions, the annual record of the hen is undoubtedly the best measure of a hen's laying capacity. Where the environmental conditions are not favorable, the short-period production, monthly or bimonthly, preferably in the season of maximum

production, gives a more certain indication of inherited egg-laying capacity than the annual record.

17. There was a close relation between rate of laying, or intensity of production, and annual production. Rate of laying is within certain limits an accurate measure of egg-laying capacity.

18. Good laying capacity was indicated (1) by heavy production at the beginning of the laying year or in the fall and early winter, (2) by heavy production or high rate of laying in any one or two months of the year, (3) by heavy production at the end of the laying year. By trap-nesting for two months at the beginning of the year, or in the spring months or in the late summer and fall, the records obtained will constitute a fairly accurate basis for culling.

19. The best two-months production is a fairly accurate basis on which to select hens of best laying capacity.

20. There was a close correlation between March and April production and annual production.

21. In the Station's experimental flocks there was a greater difference in the number of eggs laid, between the best and poorest layers in March and April than in November and December, but the percentage deviation from the mean was greater in November and December than in March and April, the months of lowest and highest production.

22. Variations in vigor and environmental conditions are the disturbing factors in the study of trap-nest records.

23. With good vigor, a hen's production may extend over a longer period than may that of a hen of the same laying capacity but without the vigor.

24. Late laying in the summer and fall did not always indicate a good layer. A hen with low rate of laying and good vigor may lay late in the summer and fall.

25. Our records show on the average a higher production in the pullet year, or first laying year, than in the second or subsequent years. There was a rather consistent decrease each year. The greatest decrease was in the production of hens with highest first-year records. Where the production was very low in the first year, there was on the average an increased production in the second year.

26. The best layers in the first year, on the average, though showing the greatest decrease in the second and third years, show better records in the second and third years than poorer layers.

27. There appears to be a correlation between rate of laying and the fat content of eggs.